AF595009

Power Electronics and Energy Management for Battery Storage Systems

Power Electronics and Energy Management for Battery Storage Systems

Editors

Andrei Blinov
Sheldon Williamson

MDPI • Basel • Beijing • Wuhan • Barcelona • Belgrade • Manchester • Tokyo • Cluj • Tianjin

Editors
Andrei Blinov
Tallinn University of Technology
Estonia

Sheldon Williamson
University of Ontario Institute of Technology
Canada

Editorial Office
MDPI
St. Alban-Anlage 66
4052 Basel, Switzerland

This is a reprint of articles from the Special Issue published online in the open access journal *Energies* (ISSN 1996-1073) (available at: https://www.mdpi.com/journal/energies/special_issues/peem_for_battery_storage_systems).

For citation purposes, cite each article independently as indicated on the article page online and as indicated below:

LastName, A.A.; LastName, B.B.; LastName, C.C. Article Title. *Journal Name* **Year**, *Volume Number*, Page Range.

ISBN 978-3-0365-5277-4 (Hbk)
ISBN 978-3-0365-5278-1 (PDF)

Contents

About the Editors

Andrei Blinov

Andrei Blinov received an M.Sc. degree in electrical drives and power electronics and a Ph.D. degree, with a dissertation devoted to the research into switching properties and performance improvement methods for high-voltage IGBT-based dc–dc converters, from Tallinn University of Technology, Tallinn, Estonia, in 2008, and 2012, respectively. After his Ph.D. studies, he spent two years in Sweden working as a Postdoctoral Researcher with the KTH Royal Institute of Technology. He is currently a Senior Researcher with the Department of Electrical Power Engineering and Mechatronics, Tallinn University of Technology. His research interests are research into switch-mode power converters, new semiconductor technologies, and energy storage systems.

Sheldon Williamson

Sheldon Williamson received his Bachelors of Engineering (B.E.) degree in Electrical Engineering with high distinction from University of Mumbai, Mumbai, India, in 1999. He received a Master's of Science (M.S.) degree in 2002, and the Doctor of Philosophy (Ph.D.) degree (with Honors) in 2006, both in Electrical Engineering, from the Illinois Institute of Technology, Chicago, IL, specializing in automotive power electronics and motor drives, at the Grainger Power Electronics and Motor Drives Laboratory. Currently, Dr. Williamson is a Professor at the Smart Transportation Electrification and Energy Research (STEER) group, within the Department of Electrical, Computer, and Software Engineering, at Ontario Tech University, in Oshawa, Ontario, Canada. He also holds the prestigious NSERC Canada Research Chair position in Electric Energy Storage Systems for Transportation Electrification. His main research interests include advanced power electronics and motor drives for transportation electrification, electric energy storage systems, and electric propulsion. Prof. Williamson is a Fellow of the IEEE.

Article

An Efficient Non-Inverting Buck-Boost Converter with Improved Step Up/Down Ability

Omar Abdel-Rahim [1,2,*], Andrii Chub [1], Andrei Blinov [1], Dmitri Vinnikov [1] and Dimosthenis Peftitsis [3]

[1] Department of Electrical Power Engineering and Mechatronics, Tallinn University of Technology, 19086 Tallinn, Estonia; andrii.chub@taltech.ee (A.C.); andrei.blinov@taltech.ee (A.B.); dmitri.vinnikov@taltech.ee (D.V.)
[2] Electrical Engineering Department, Aswan University, Aswan 81542, Egypt
[3] Department of Electrical Power Engineering, Norwegian University of Science and Technology, NO-7491 Trondheim, Norway; dimosthenis.peftitsis@ntnu.no
* Correspondence: o.abdelrahim@aswu.edu.eg

Abstract: In this article, a new non-inverting buck-boost converter with superior characteristics in both bucking and boosting is presented. The proposed converter has some distinct features, such as high step-up/-down ability and low voltage/current stress on its switching devices. The voltage gain of the proposed converter is double the reported value for the traditional buck-boost converter. Although it has three switches, the three switches operate simultaneously, hence no dead-time is required. Two out of the three switches are under voltage stress equal to half of the output voltage. The overall efficiency of the system is promising because of the ability to select devices with low voltage drops. Converter analysis and steady-state performance in both continuous conduction mode (CCM) and discontinuous conduction mode (DCM) are presented in detail. A 1 kW hardware prototype of the converter was implemented in the laboratory; with a step-up ratio of 3.5 and 1 kW power, the measured efficiency is above 95.4%, and with step-up ratio 8, it is around 91.5%.

Keywords: high-gain non-inverting buck-boost converter; continuous conduction mode (CCM); discontinuous conduction mode (DCM)

Citation: Abdel-Rahim, O.; Chub, A.; Blinov, A.; Vinnikov, D.; Peftitsis, D. An Efficient Non-Inverting Buck-Boost Converter with Improved Step Up/Down Ability. *Energies* **2022**, *15*, 4550. https://doi.org/10.3390/en15134550

Academic Editors: Alon Kuperman and Marco Pau

Received: 17 May 2022
Accepted: 20 June 2022
Published: 22 June 2022

1. Introduction

Traditional buck-boost converters, CUK and SEPIC, are able to buck or boost input voltage; however, their bucking or boosting abilities are limited, and they have high stress on their switching devices, hence their efficiency and applications are limited [1–6]. In order to improve step-up/-down abilities, a group of power converters have been developed in the literature [7–17]. The topology proposed in [7] is a modification of the traditional buck-boost converter with improved voltage gain, but it has an inverted output and two of the switching devices are under high voltage stress. A high gain with continuous input current buck-boost converter has been proposed in [7,8], but the converter is inverted and includes many storage devices. In [9], a novel buck-boost converter is proposed with lower component stresses and less storage devices, However, the converter has limited voltage gain; high ripple; and the converter switches operate in a complementary manner, which increases dead-time and switching protection issues.

The quadratic voltage gain buck-boost converters developed in [10–12] provide good performance in step-up mode, but their step-down ability is very limited.

In [13,14], semi-quadratic buck-boost converters are proposed. Despite their improved performance in both bucking and boosting modes, there is no common ground and the input current is discontinuous. A quasi-Y source-based buck-boost dc–dc converter is introduced in [15,16]. This converter achieved a very high voltage gain using two inductors. Nevertheless, the severe slope of the voltage gain ratio makes controlling the converter very difficult.

In order to achieve higher voltage and gain and sustain higher efficiency at a wide range of input voltage change, this paper presents a new high-gain non-inverting buck-boost converter. The structure proposed has different merits, such as non-inverting, high voltage gain, reduced components' stresses, and the ability to sustain better efficiency at wide voltage and load ranges.

The rest of the paper is organized as follows: Section 2 discusses the principle of operation and analysis of the proposed converter; Section 3 presents the experimental results of the converter; and finally, Section 4 presents the conclusions.

2. Proposed Buck-Boost Dc–Dc Converter

The configuration of the proposed buck-boost dc–dc converter is illustrated in Figure 1 [17]. The structure is implemented using three power switches (S_1, S_2, S_3), two diodes (D_1, D_o), two inductors (L_1, L_2), and an output capacitor (C_o). The three switches are triggered on and off simultaneously, and the diodes operate as freewheeling diodes. The two inductors charge in parallel and discharge in series.

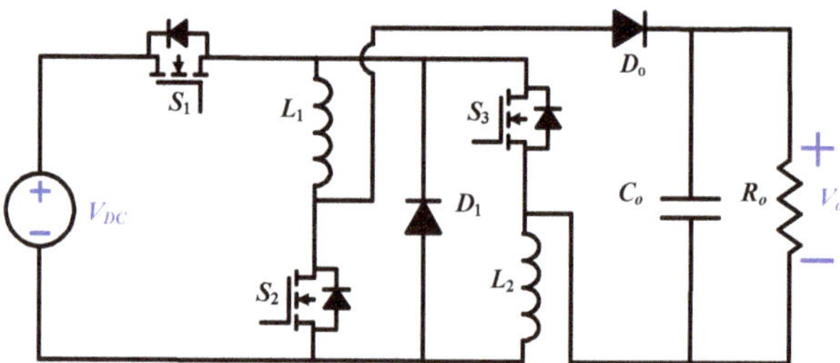

Figure 1. Proposed buck-boost dc–dc converter.

The converter is able to operate in continuous conduction mode (CCM) and discontinuous conduction mode (DCM). Both modes of operation will be considered in the following sections.

2.1. Continuous Conduction Mode

In order to simplify the analysis of the CCM mode, two assumptions are considered in the forthcoming analysis:

- ✓ Capacitor voltage ripple is very small compared with the voltage itself, thus it could be neglected.
- ✓ Inductor current ripple is negligible because of its very small value.
- ✓ All semiconductor devices are ideal.

The converter power switches are triggered ON and OFF simultaneously, hence the converter will have two operating modes; see Figure 2a,b. Typical waveforms of the converter in CCM are shown in Figure 3.

Mode I [0-DT_S]: In this time period, switches (S_1, S_2, S_3) are turned ON, while diodes (D_1, D_o) are turned OFF. This mode is illustrated in Figure 2a. As can be seen from the figure, the two inductors charge in parallel from the source. Applying Kirchhoff voltage law (KVL) and Kirchhoff current law (KCL) to Figure 2a, the following equations are deduced:

$$v_{L1} = v_{L2} = V_{dc} \tag{1}$$

$$i_c = -\frac{V_o}{R} \tag{2}$$

$$i_d = 0 \tag{3}$$

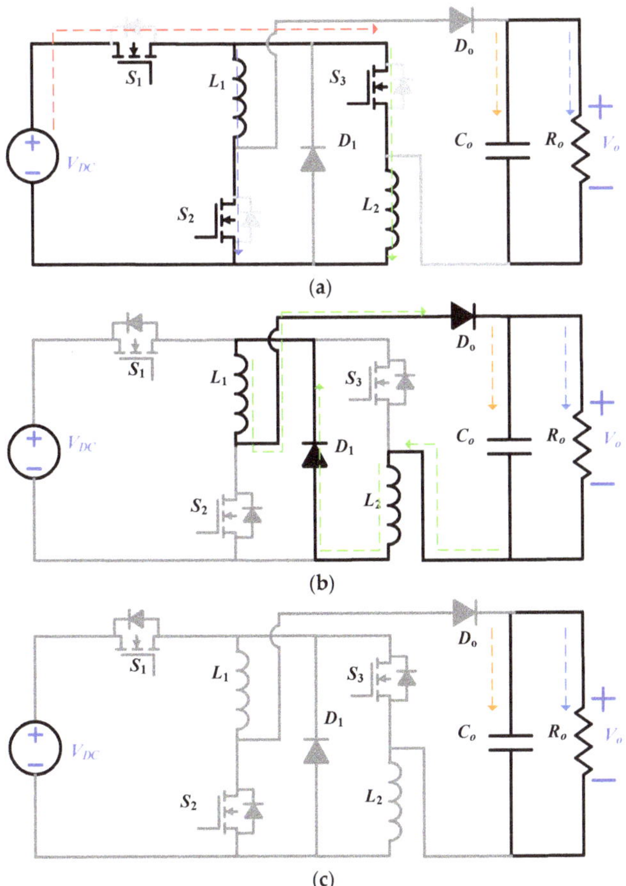

Figure 2. Operation modes of the proposed converter: (**a**) operation mode #1, (**b**) operation mode #2, and (**c**) operation mode #3.

Mode II [DT_S-T_S]: In this time period, switches (S_1, S_2, S_3) are turned OFF and, consequently, diodes (D_1, D_o) are turned ON to provide a freewheeling path for the current. This mode is illustrated in Figure 2b. As can be investigated from the figure, the two inductors discharge their energies to the load in series. Applying Kirchhoff voltage law (KVL) and Kirchhoff current law (KCL) to Figure 2b, the following equations are deduced:

$$v_L = v_{L1} = v_{L2} \tag{4}$$

$$2v_L = -V_o \tag{5}$$

$$i_C = I_L - \frac{V_o}{R} \tag{6}$$

$$i_d = I_L \tag{7}$$

The steady-state voltage gain of the proposed converter could be deduced from the analysis of the two modes of operation by applying voltage second balance, and the voltage gain of the proposed converter is as follows:

$$\frac{V_o}{V_{dc}} = M = \frac{2D}{(1-D)} \tag{8}$$

where v_{L1}, v_{L2}, V_{dc}, V_o, I_L, i_C, i_d, M, and D are inductor L_1 voltage, inductor L_2 voltage, input voltage, output voltage, inductor current, capacitor current, diode current, voltage gain, and duty cycle, respectively.

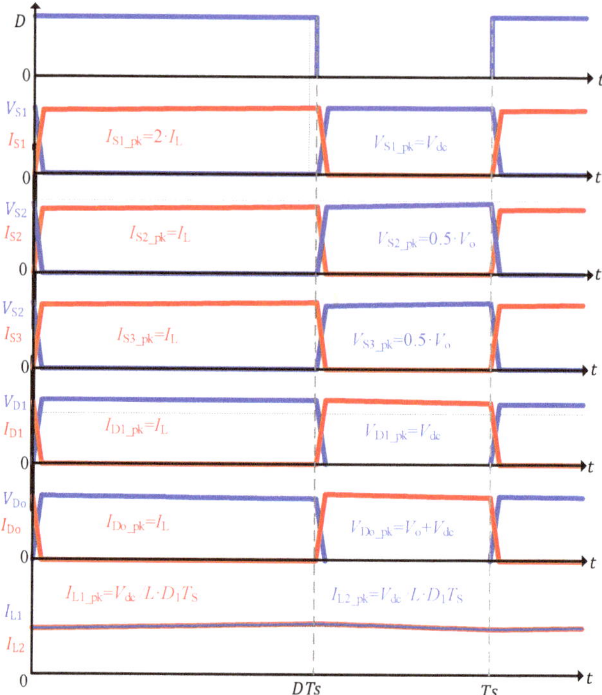

Figure 3. Typical converter waveforms in CCM.

2.2. Discontinuous Conduction Mode

The discontinuous conduction mode typically occurs with large inductor current ripple in a converter operating at light load and containing current unidirectional switches. However, some converters are purposely designed to operate in DCM. The proposed converter will have three modes of operation while operating in DCM; see Figure 2a–c. The typical converter waveform while operating in DCM is illustrated in Figure 4.

Mode I and Mode II, which were discussed in the previous section, are similar to CCM analysis.

Mode III: In this interval, both the power switches and diodes are turned off. The inductors' currents are zero, as illustrated in Figure 2c.

$$v_L = 0 \tag{9}$$

$$i_C - \frac{V_o}{R} \tag{10}$$

$$i_d = 0 \tag{11}$$

Applying inductor volt-second balance in Equations (1), (5), and (9), the relation between input and output voltage is obtained:

$$D_1 * v_{dc} = D_2 * v_o \tag{12}$$

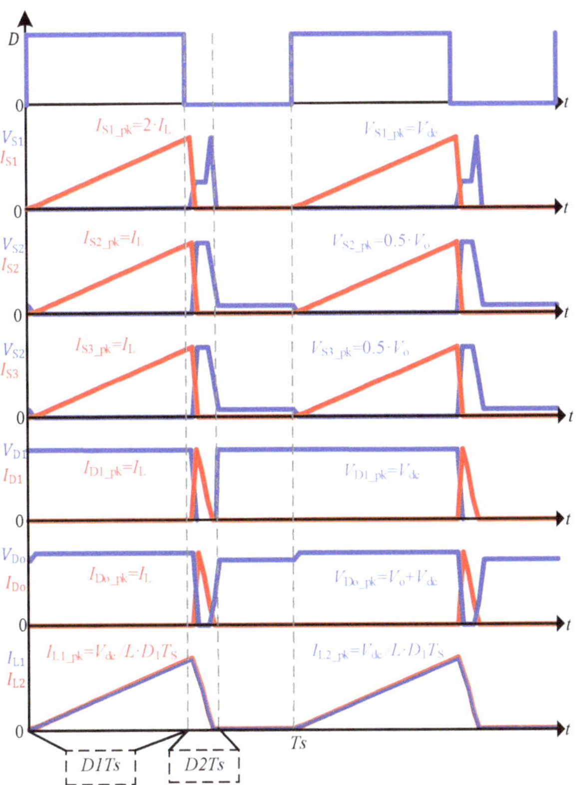

Figure 4. Typical converter waveforms in DCM.

The duty cycle D_2 is an unknown, so a second equation is needed to eliminate D_2. Capacitor charge-balance is used to obtain the second equation. The average of the diode current is equal to the output current:

$$\langle i_d \rangle = \frac{V_o}{R} \tag{13}$$

A sketch of the inductor and diode currents in DCM is illustrated in Figure 5a,b. The dc component of the diode current is given by

$$\langle i_d \rangle = \frac{1}{T_S} \int_0^{T_S} i_d(t) dt \tag{14}$$

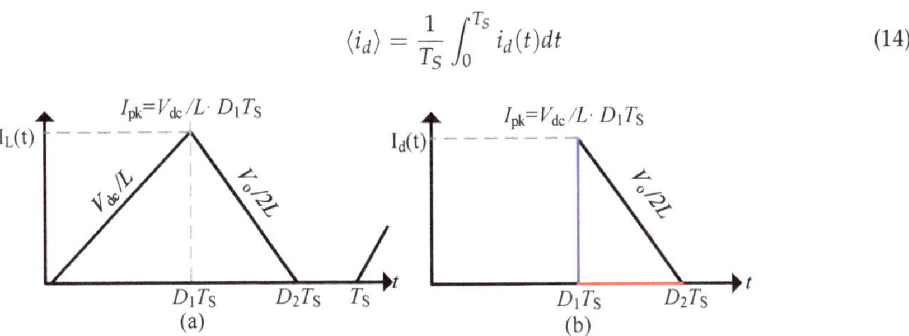

Figure 5. DCM operation (**a**) inductor current and (**b**) diode current.

The peak diode current could be obtained from the graph as

$$i_{pk} = \frac{V_o}{2L} * D_1 * T_S \tag{15}$$

Solving Equations (13)–(15), the second required equation is obtained as

$$\langle i_d \rangle = \frac{1}{2} * D_2 * T_S * \frac{V_o}{2L} * D_1 = \frac{V_o}{R} \tag{16}$$

Let

$$k = \frac{2L}{RT_S} \tag{17}$$

Then

$$D_2 = 2 * k / D_1 \tag{18}$$

Finally, the converter voltage gain in DCM operation is given as

$$\frac{V_o}{V_{dc}} = M = \frac{D_1^2}{(2 * K)} \tag{19}$$

where D_1, D_2, T_S, and R_0 are periods when the switches are conducting, periods when the diode is conducting, switching time, and load resistance, respectively.

The boundary for CCM and DC operation can be obtained by relating inductor current and inductor ripple

$$I_L > \Delta i_L \text{ For CCM} \tag{20}$$

$$I_L < \Delta i_L \text{ For DCM} \tag{21}$$

Substituting CCM solutions for I_L and Δi_L in (20)

$$\frac{V_{dc}}{R} * \left(\frac{2D}{1-D}\right)^2 > \frac{V_{dc}}{2L} * DT_S \tag{22}$$

Equation (22) could be rearranged to

$$\frac{2L}{RT_S} = K > D * \left(\frac{2D}{1-D}\right)^2 \tag{23}$$

Hence

$$K_{cri} = D * \left(\frac{2D}{1-D}\right)^2 \tag{24}$$

where K_{cri} is the critical boundary between CCM and DCM.

According to the above analysis, the converter can operate on CCM or DCM based on the operating conditions; in order to avoid such conditions, accurate design of the converter must be considered. Figure 6 represents the boundary condition between CCM and DCM at different duty cycles and different power while the output voltage is fixed at 350 V.

2.3. Switches' and Diodes' Voltage Stresses

Voltage and current stress are important parameters in designing and selecting circuit parameters, and the proposed converter switching elements' stress is discussed below.

Switches S_1, S_2, and S_3 are triggered in a simultaneous manner, but their ratings are different. The voltage stress of switch S_1 is equal to

$$V_{dsS1} = V_{dc} \tag{25}$$

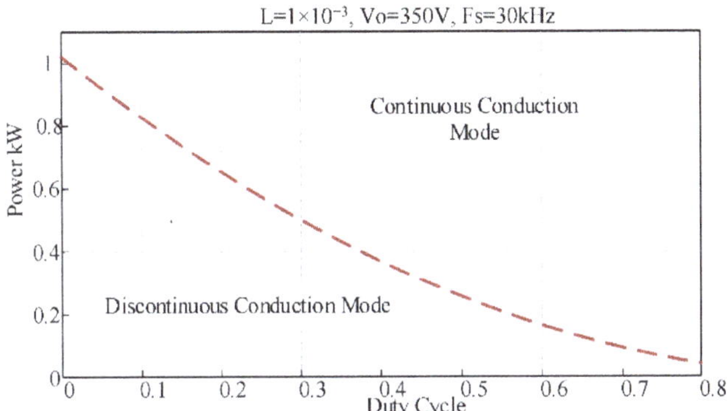

Figure 6. CCM and DC boundary.

The current stress of switch S_1 is given by

$$I_{S1} = 2 * I_L \tag{26}$$

Switches S_2 and S_3 face similar voltage and current stress, as follows:

$$V_{dsS2} = V_{dsS3} = \frac{V_o}{2} \tag{27}$$

$$I_{S2} = I_{S3} = I_L \tag{28}$$

Diodes D_1 and D_o work as freewheeling diodes and are activated in complementary manners to the switches. The voltage and current stress of both diodes are given by

$$V_{d1} = V_{dc} \tag{29}$$

$$I_{d1} = I_L \tag{30}$$

$$V_{do} = V_{dc} + V_o \tag{31}$$

$$I_{S2} = I_L \tag{32}$$

A depiction of the devices' normalized voltage stresses with different voltage gain is illustrated in Figure 7. In Figure 7, the voltage stress is normalized to the input voltage.

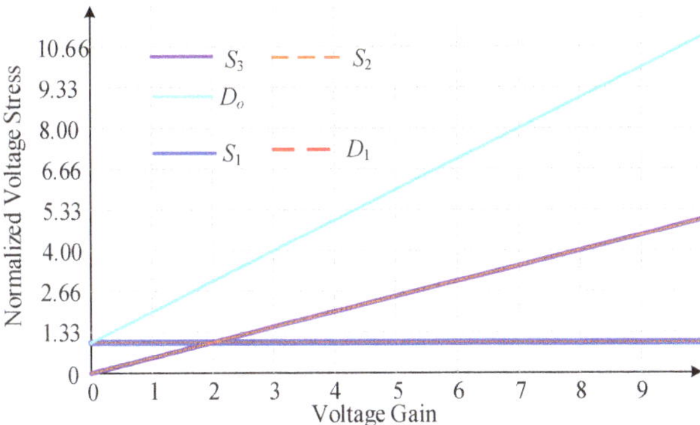

Figure 7. Circuit component voltage stress normalized to input voltage vs. converter voltage gain.

2.4. Components' Design

The design of the circuit parameters, inductors, and capacitor is obtained from the steady-state analysis performed in the previous sections. Utilizing inductor volt-second balance and capacitor charge, the designs of parameters are as follows:

2.4.1. Inductors' Design

Inductors' selection is based on the required ripple of its current. The inductor current ripple in CCM is drawn in Figure 8a and is given by

$$\Delta i_L = \Delta i_L^{on} = \Delta i_L^{off} = \frac{V_{dc} D T_S}{L} = \frac{V_o(1-D)T_S}{2L} \quad (33)$$

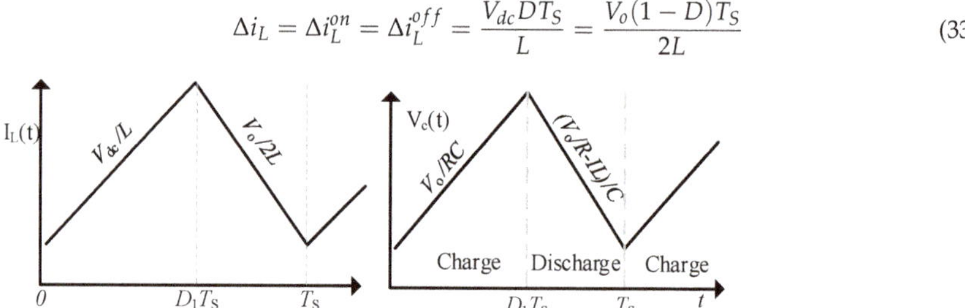

Figure 8. CCM operation (**a**) inductor current and (**b**) output capacitor voltage.

This equation is valid in both CCM and DCM. By defining the required amount of ripple, the inductor value could be defined as follows:

$$L = \frac{(V_o * (1-D)) * T_S}{(2 * \Delta i_L)} \quad (34)$$

Based on Equation (29), there is a dependency between inductance L and duty cycle D. In order to avoid any misoperation of the converter, we design the inductance based on the extreme condition that the current ripple at the extreme scenario does not exceed the required ripple and when duty cycle below the ripple will be below the required level.

Let us assume the required ripple Δi_L is 10%, then we can calculate L at the duty cycle around 0.82. Then, when the duty cycle is lower than 0.85, the ripple will be less than 10%.

2.4.2. Capacitors' Design

The output capacitor value is selected based on the amount of voltage ripple acceptable in the output voltage. The output capacitor voltage waveform is illustrated in Figure 8b and the ripple equation is as follows:

$$\Delta v = \Delta v_c^{on} = \Delta v_o^{off} = \frac{(V_o - IL)(1-D)T_S}{RC} = \frac{V_o D T_S}{RC} \quad (35)$$

This equation is valid in both CCM and DCM. By defining the required amount of ripple, the capacitor value could be defined as follows:

$$C_o = \frac{(V_o * D * T_S)}{(\Delta v * R)} \quad (36)$$

where Δi_L, Δi_L^{on}, Δi_L^{off}, L, Δv, Δv_c^{on}, Δv_o^{off}, and C are inductor current ripple, inductor ripple while the inductor is charging, inductor ripple while the inductor is discharging, inductor value, capacitor ripple, capacitor ripple while the capacitor is charging, capacitor ripple while the capacitor is discharging, and capacitor value, respectively.

Based on Equation (30), there is a dependency between capacitance C and duty cycle D. In order to avoid any misoperation of the converter, we design the capacitance based on the extreme condition that the voltage ripple at the extreme scenario does not exceed the required ripple and when duty cycle below the ripple will be below the required level

Let us assume the required ripple Δv is 10%, then we can calculate C at the duty cycle around 0.82. Then, when the duty cycle is lower than 0.85, the ripple will be less than 10%.

2.4.3. Comparative Study

Converter performance mainly depends on the input voltage, input power, and step-up ratio. The converter voltage gain is affected by the loading profile and supply voltage. In Figure 9a, the input voltage is fixed at 150 V, while different loading profiles are applied; at light load, the converter conversion ability is higher than at heavy loading. The second case study is illustrated in Figure 9b, where load profile is fixed with different supply voltages; as the supply voltage increases, the step-up/-down ability increases.

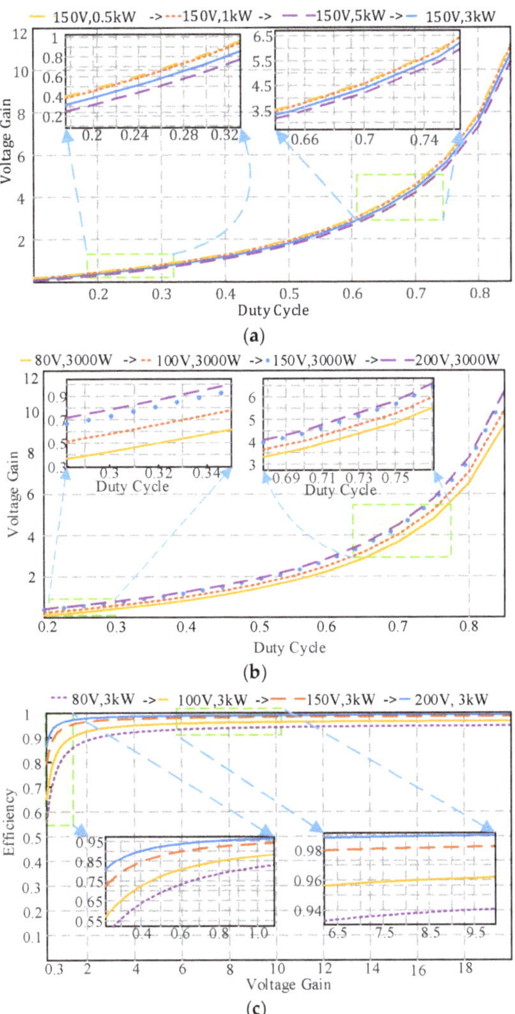

Figure 9. Different cases of study for the proposed converter: (**a**) Voltage gain at different duty cycle and different loading. (**b**) Voltage gain with fixed loading and different input voltages. (**c**) Converter efficiency at different input voltages.

However, in both case studies, the differences in the voltage gain do not have a very high ratio.

Converter efficiency depends on many factors such as the load profile, source voltage, and voltage gain. In the scenario illustrated in Figure 9c, the load profile is fixed while both voltage gain and source voltage are variable. In buck mode, as the source voltage increases and the bucking ratio is lowered, the converter demonstrates the highest efficiency, while with lower input voltage and a higher bucking ratio, the converter efficiency is low. During boosting mode, as source voltage increases, efficiency increases too.

Another case study is considered in Figure 10a, where input voltage is set to 150 V, while load profile is variable and efficiency is measured at different voltage gains. With heavy loading, the converter demonstrates lower efficiency than with a medium or moderate loading profile. A comparison between the proposed converter and different converters reported in the literature is illustrated in Table 1. In the voltage gain comparison illustrated in Figure 10b, both the proposed converter and switched inductor buck-boost converter have similar step-up/-down ability, but the proposed converter has higher efficiency; see Figure 10c.

Table 1. Voltage gain and component stress comparison.

Converter Topology	Gain M = V_o/Vin	Components' Count				Switches' and Diodes' Voltage Stress	Switches' and Diodes' Current Stress
		Switch	Diode	L	C		
Buck-Boost [18]	$(D/(1\text{-}D))$	1	1	1	1	$S{:}V_o$ $D{:}(V_o+V_{in})$	I_L I_L
Non-Inverting [19]	$(D/(1\text{-}D))$	2	2	1	1	$S_1{:}V_{in}$ $S_2{:}V_o$ $D_1{:}V_{in}$ $D_o{:}V_o$	I_L I_L I_L I_L
Cuk [5]	$(D/(1\text{-}D))$	1	2	2	1	$S_1{:}V_{in}/(1\text{-}D)$ $D_1{:}V_{in}$ $D_o{:}V_o$	I_L I_L I_L
SEPIC [19]	$(D/(1\text{-}D))$	1	1	2	2	$S_1{:}V_o+V_{in}$ $D_1{:}V_o+V_{in}$	$2I_L$ I_L
SIBBC [20]	$2D/(1\text{-}D)$	1	4	2	1	$S_1{:}V_o+V_{in}$ $D_1{:}V_{in}$ $D_2{:}V_o/2$ $D_3{:}V_o/2$ $D_o{:}V_o+V_{in}$	$2I_L$ I_L I_L I_L I_L
Lakshmi [21]	$(1+D1)/(1\text{-}D_1\text{-}D_2)$	3	2	2	1	$S_1{:}(V_o+V_{in})/2$ $S_2{:}(V_o+V_{in})/2$ $S_3{:}V_o$ $D_1{:}V_o$ $D_o{:}V_o+V_{in}$	I_L I_L I_L I_L I_L
[22]	$2D/(1\text{-}D)^2$	2	3	2	3	$S_1{:}1/(1\text{-}D)^*V_{in}$ $S_2{:}V_{in}{}^*(1+D)/(1\text{-}D)^2$ $D_1{:}V_{in}/(1\text{-}D)$ $D_2{:}V_{in}/(1\text{-}D)$ $D_o{:}V_o+V_{in}/(1\text{-}D)$	I_L I_L I_L I_L I_L
[23]	$D^2/(1\text{-}D)^2$	2	2	2	2	$S_1{:}1/(1\text{-}D)^*V_{in}$ $S_2{:}V_{in}{}^*D/(1\text{-}D)^2$ $D_1{:}V_{in}/(1\text{-}D)$ $D_2{:}V_{in}{}^*D/(1\text{-}D)^2$	$2I_L$ I_L I_L I_L

Table 1. *Cont.*

Converter Topology	Gain $M = V_o/Vin$	Components' Count				Switches' and Diodes' Voltage Stress	Switches' and Diodes' Current Stress
		Switch	Diode	L	C		
Proposed	$2D/(1\text{-}D)$	3	2	2	1	$S_1{:}V_{in}$ $S_2{:}V_o/2$ $S_3{:}V_o/2$ $D_1{:}V_{in}$ $D_o{:}V_o{+}V_{in}$	$2I_L$ I_L I_L I_L I_L

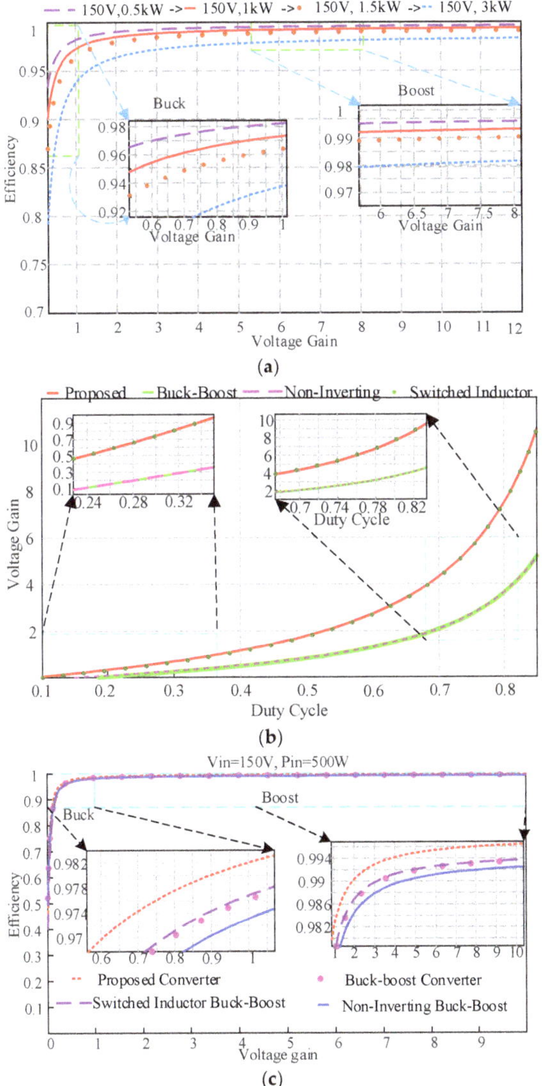

Figure 10. (**a**) Converter efficiency at fixed input voltage and different loading, (**b**) voltage gain comparison among the proposed and other buck-boost converters, and (**c**) efficiency comparison among the proposed and other buck-boost converters.

3. Experimental Verification

This section provides the experimental results of the developed system, and the parameters used to build the prototype are illustrated in Table 2. A photo of the proposed system is shown in Figure 11.

Table 2. Hardware prototype specifications.

Parameter	Value
Input voltage range [*V*]	33–150
P [*W*]	700 W
F_s *Switching Frequency*	30 kHz
Switches S_1, S_2, S_3	IMZ120R030M1HXKSA1
Diodes D_1, D_o	DPG10I300PA
Inductors $L_1 = L_2$	1 mH
Capacitor C_o	320 μF

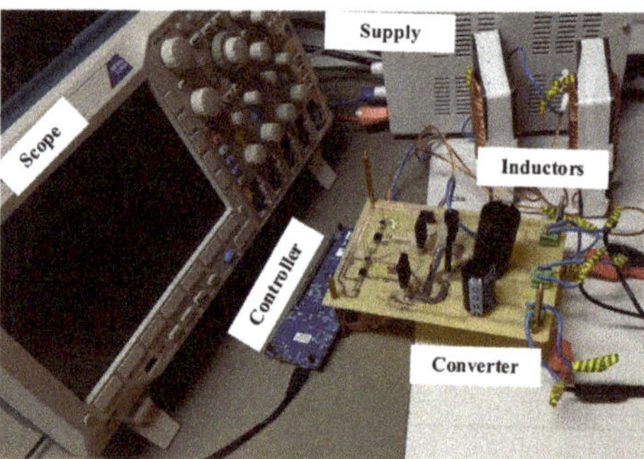

Figure 11. Experimental set-up schematic.

A case study where the duty cycle is set to 0.6 with 30 V input voltage is illustrated in Figure 12. Three switches are operating in synchronous manner, hence the gate source pulses for the three switches are the same as illustrated in Figure 12a. Diodes D_1 and D_o are operating as freewheeling diodes. The cathode–anode voltages of the two diodes are illustrated in Figure 12b. The input current is the sum of the two inductors' currents when the switches are on and zero when the switches are off, and the input capacitor smoothens the input current. The drain source voltages of the three switches are illustrated in Figure 12c. Switches S2 and S3 face the same voltage stress and carry the same current.

Figure 12d illustrates switch S_1 current, which is equal to the sum of the two inductors' currents. Switch S_2 current is illustrated in Figure 12e, which is equal to the inductor current. The output diode current is illustrated in Figure 12f, where spikes are noted in the switches and diode currents because of a problem in the used probe; however, it does not exist in real current as there are no spikes in the measured voltages.

A boosting case study is considered in Figure 13, where the input voltage is 25 V and the output voltage generated is around 38 V, and a bucking case study is illustrated in Figure 14, where the input voltage is 23 V, output voltage is 9.25 V, duty cycle is 0.2244, and voltage gain is 0.4.

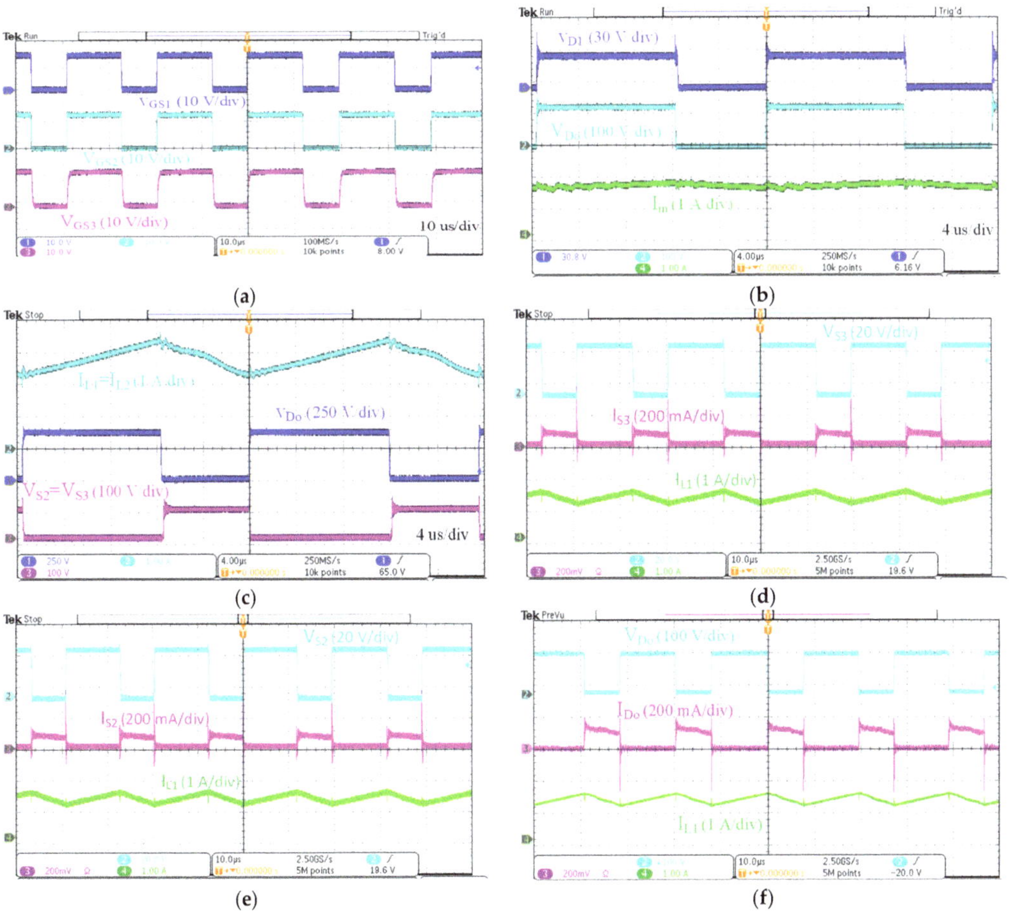

Figure 12. Experimental results of converter at duty cycle of 0.6 and input voltage of 30 V: (**a**) gate source pulses; (**b**) Ch1 diode D_1 voltage, diode D_o voltage, and input current; (**c**) Ch1 output diode D_o voltage, Ch3 switches' S_2 and S_3 voltage, and Ch2 inductor L_1 current; (**d**) Ch1 switch S_3 voltage; Ch2 switch S_3 currents; (**e**) Ch2 switch S_2 voltage, Ch3 switch S_2 current, and inductor current; and (**f**) Ch2 diode D_o voltage and diode D_o current.

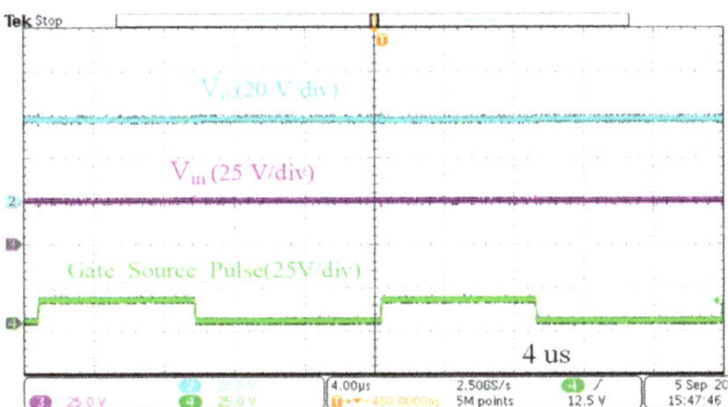

Figure 13. Boosting case study, where the input voltage is 25 V, output voltage is 38 V, duty cycle is 0.428, and voltage gain is 1.52.

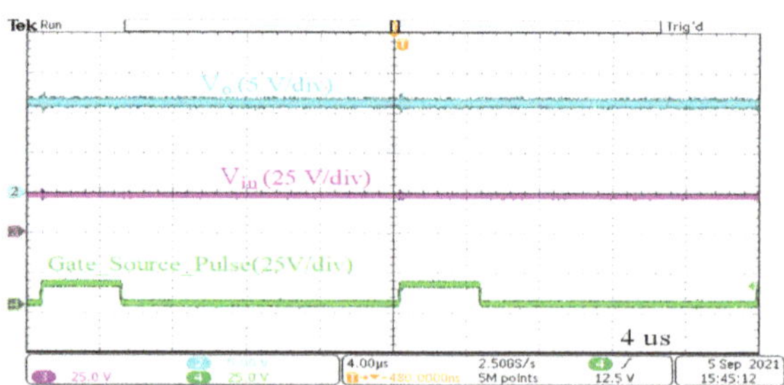

Figure 14. Bucking case study, where the input voltage is 23 V, output voltage is 9.25 V, duty cycle is 0.2244, and voltage gain is 0.4.

The converter voltage gain was measured experimentally, and the theoretical and measured voltage gains of the converter with varying duty cycles are illustrated in Figure 15. For comparison purposes, three prototypes were built in the laboratory for the traditional buck-boost, non-inverting buck-boost, and proposed converter. The three prototypes were built using the same parameters as in Table 2. In the first case study, which is illustrated in Figure 16, the input voltage is set to 100 V and the step ratio is fixed at 3.7. For such a step-up ratio, the proposed converter requires a duty cycle of 0.68, while the conventional and non-inverting buck-boost converters both require a duty cycle of 0.8.

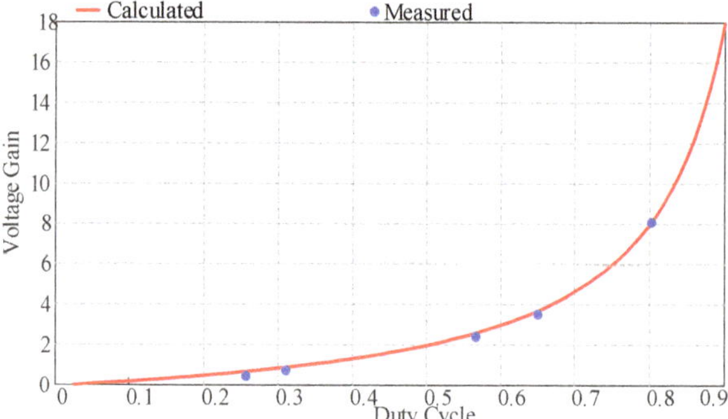

Figure 15. Calculated and measured converter voltage gain vs. duty cycle.

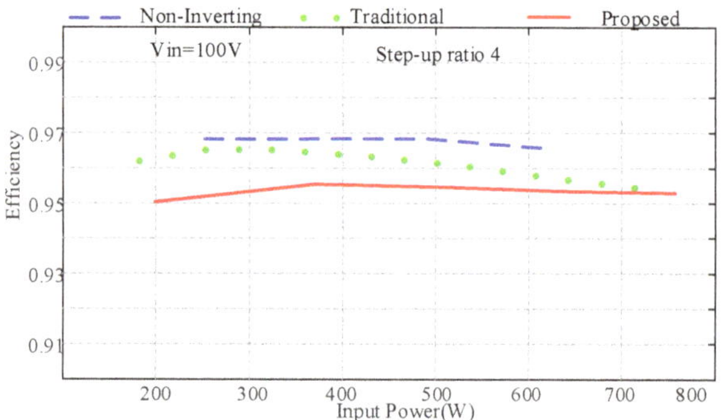

Figure 16. Measured efficiency comparison between the proposed converter, non-inverting buck-boost, and traditional buck-boost converter at a step-up ratio of 3.7.

In this case study, the non-inverting converter demonstrates the highest efficiency, while at high power, both the proposed and conventional converter have the same efficiency. In the second case study, which is illustrated in Figure 17, the input voltage is fixed at 30 V and the step-up ratio is 8. The efficiency of the proposed converter and the non-inverting converter is comparable, but with the increase in power (over 300 W), the proposed converter demonstrates the highest efficiency.

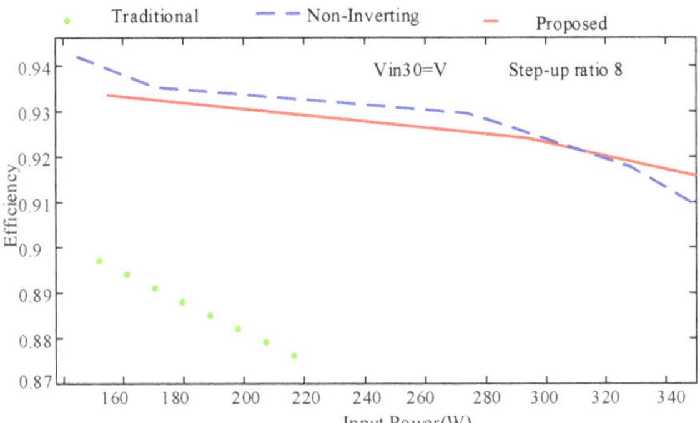

Figure 17. Measured efficiency comparison between the proposed, non-inverting, and traditional buck-boost converter at a step-up ratio of 8.

The last case study demonstrates step-down comparison. In Figure 18, the input voltage is fixed at 150 V and the step-down ratio is 3. The non-inverting converter demonstrates the lowest efficiency. The proposed converter and the conventional converter demonstrate comparable efficiency at low power, but with the increase in input power, the proposed converter demonstrates the highest efficiency.

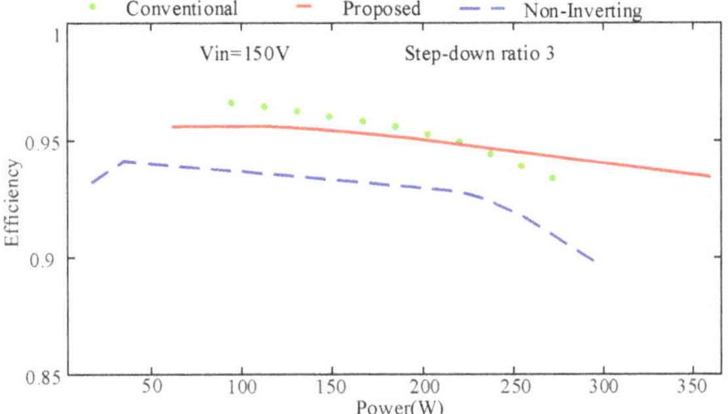

Figure 18. Measured efficiency comparison between the proposed, non-inverting buck-boost, and traditional buck-boost converter at a step-down ratio of 3.

4. Conclusions

In this paper, a new non-inverting high-gain buck-boost structure is developed with improved step-up/-down ability. The performance of the converter in both CCM and DCM is studied and analyzed. The design of the converter elements is investigated and described. The operating conditions and voltage/current stress of each device are studied. Based on the performed analysis, the proposed converter devices are under low voltage and current stress compared with other buck-boost converters. A 700 W prototype was built for the converter to investigate its performance experimentally. The efficiency of the proposed converter is measured at different voltage gains and compared with the traditional buck-boost converter.

The theoretical and measured voltage gain matched. While working in step-up, the converter demonstrated better performance at high power. The peak measured efficiency of the converter at a step-up ratio of 3.7 was 95.4%.

Author Contributions: O.A.-R. developed and simulated the idea; O.A.-R., A.C. and A.B. implemented the idea prototype and verified the experimental results; O.A.-R. and D.V. revised the data and results; O.A.-R. wrote the first draft; O.A.-R., D.V. and D.P. revised the manuscript. All authors have read and agreed to the published version of the manuscript.

Funding: This research was financed in part by the European Economic Area (EEA) and Norway Financial Mechanism 2014–2021 under Grant EMP474 and in part by the Estonian Research Council under Grant PRG1086 and the programme Mobilitas Pluss (Grant MOBJD1033).

Conflicts of Interest: The authors declare no conflict of interest.

References

1. Abdel-Rahim, O.; Chub, A.; Blinov, A.; Vinnikov, D. Buck-Boost Resonant Z-Source Parital Power Converter. In Proceedings of the 2022 3rd International Conference on Smart Grid and Renewable Energy (SGRE), Doha, Qatar, 20–22 March 2022; pp. 1–6. [CrossRef]
2. Abdel-Rahim, O.; Orabi, M.; Ahmed, M.E. Buck-Boost Interleaved Inverter For Grid Connected Photovoltaic System. In Proceedings of the 2010 IEEE International Conference on Power and Energy, Kuala Lumpur, Malaysia, 29 November–1 December 2010; pp. 63–68. [CrossRef]
3. Callegaro, L.; Ciobotaru, M.; Fletcher, J.E. An Intelligent Pass-Through Algorithm for Non-Inverting Buck-Boost Solar Power Optimizers. In Proceedings of the 2019 21st European Conference on Power Electronics and Applications (EPE '19 ECCE Europe), Genova, Italy, 3–5 September 2019; pp. P.1–P.10. [CrossRef]
4. Linares, L.; Erickson, R.W.; MacAlpine, S.; Brandemuehl, M. Improved Energy Capture in Series String Photovoltaics via Smart Distributed Power Electronics. In Proceedings of the 2009 Twenty-Fourth Annual IEEE Applied Power Electronics Conference and Exposition, Washington, DC, USA, 15–19 February 2009; pp. 904–910. [CrossRef]
5. Abdel-Rahim, O.; Orabi, M.; Ahmed, M.E. High Gain Single-Stage Inverter For Photovoltaic AC Modules. In Proceedings of the 2011 Twenty-Sixth Annual IEEE Applied Power Electronics Conference and Exposition (APEC), Fort Worth, TX, USA, 6–11 March 2011; pp. 1961–1967. [CrossRef]
6. Banaei, M.R.; Bonab, H.A.F. A High Efficiency Nonisolated Buck–Boost Converter Based on ZETA Converter. *IEEE Trans. Ind. Electron.* **2019**, *67*, 1991–1998. [CrossRef]
7. Banaei, M.R.; Bonab, H.A.F. A novel structure for single-switch nonisolated transformerless buck–boost DC–DC con-verte. *IEEE Trans. Ind. Electron.* **2016**, *64*, 198–205. [CrossRef]
8. Ajami, A.; Ardi, H.; Farakhor, A. Design, analysis and implementation of a buck–boost DC/DC converter. *IET Power Electron.* **2014**, *7*, 2902–2913. [CrossRef]
9. Son, H.-S.; Kim, J.-K.; Lee, J.-B.; Moon, S.-S.; Park, J.-H.; Lee, S.-H. A New Buck–Boost Converter With Low-Voltage Stress and Reduced Conducting Components. *IEEE Trans. Ind. Electron.* **2017**, *64*, 7030–7038. [CrossRef]
10. Abdel-Rahim, O.; Furiato, H. Switched Inductor Quadratic Boosting Ratio Inverter With Proportional Resonant Controller For Grid-Tie PV Applications. In Proceedings of the IECON 2014—40th Annual Conference of the IEEE Industrial Electronics Society, Dallas, TX, USA, 29 October–1 November 2014; pp. 5606–5611. [CrossRef]
11. Maldonado, J.C.M.; Valdez-Resendiz, J.E.E.; Vite, P.M.G.; Rosas-Caro, J.C.; Rivera-Espinosa, M.D.R.; Valderrabano-Gonzalez, A. Quadratic Buck–Boost Converter with Zero Output Voltage Ripple at a Selectable Operating Point. *IEEE Trans. Ind. Appl.* **2019**, *55*, 2813–2822. [CrossRef]
12. Rosas-Caro, J.C.; Valdez-Resendiz, J.E.; Mayo-Maldonado, J.C.; Alejo-Reyes, A.; Valderrabano-Gonzalez, A. Quadratic buck–boost converter with positive output voltage and minimum ripple point design. *IET Power Electron.* **2018**, *11*, 1306–1313. [CrossRef]
13. Nousiainen, L.; Suntio, T. Dynamic characteristics of current-fed semi-quadratic buck-boost converter in photovoltaic applications. In Proceedings of the 2011 IEEE Energy Conversion Congress and Exposition, Phoenix, AZ, USA, 17–22 September 2011; pp. 1031–1038. [CrossRef]
14. Yari, K.; Shahalami, S.H.; Mojallali, H. A Novel Nonisolated Buck–Boost Converter with Continuous Input Current and Semiquadratic Voltage Gain. *IEEE J. Emerg. Sel. Top. Power Electron.* **2021**, *9*, 6124–6138. [CrossRef]
15. Yari, K.; Mojallali, H.; Shahalami, S.H. A New Coupled-Inductor-Based Buck–Boost DC–DC Converter for PV Applications. *IEEE Trans. Power Electron.* **2021**, *37*, 687–699. [CrossRef]
16. AL-Hakeem, K.M.; Ahmad, O.A. Buck-Boost In-Out of Phase Y-source AC Matrix Converter. In Proceedings of the 2018 Third Scientific Conference of Electrical Engineering (SCEE), Baghdad, Iraq, 19–20 December 2018; pp. 274–279. [CrossRef]
17. Abdel-Rahim, O.; Chub, A.; Blinov, A.; Vinnikov, D. New High-Gain Non-Inverting Buck-Boost Converter. In Proceedings of the IECON 2021—47th Annual Conference of the IEEE Industrial Electronics Society, Toronto, ON, Canada, 13–16 October 2021; pp. 1–6. [CrossRef]
18. Smedley, K.; Cuk, S. Dynamics of one-cycle controlled Cuk converters. *IEEE Trans. Power Electron.* **1995**, *10*, 634–639. [CrossRef]

19. Ericson, R.; Maksimovic, D. *Fundamentals of Power Electronics*; Kluwer Academic Publishers: Dordrecht, The Netherlands, 2007.
20. Abdel-Rahim, O.; Orabi, M.; Abdelkarim, E.; Ahmed, M.; Youssef, M.Z. Switched Inductor Boost Converter For PV Applications. In Proceedings of the 2012 Twenty-Seventh Annual IEEE Applied Power Electronics Conference and Exposition (APEC), Orlando, FL, USA, 5–9 February 2012; pp. 2100–2106. [CrossRef]
21. Lakshmi, M.; Hemamalini, S. Nonisolated High Gain DC–DC Converter for DC Microgrids. *IEEE Trans. Ind. Electron.* **2017**, *65*, 1205–1212. [CrossRef]
22. Sarikhani, A.; Allahverdinejad, B.; Hamzeh, M. A Nonisolated Buck–Boost DC–DC Converter with Continuous Input Current for Photovoltaic Applications. *IEEE J. Emerg. Sel. Top. Power Electron.* **2020**, *9*, 804–811. [CrossRef]
23. Miao, S.; Wang, F.; Ma, X. A New Transformerless Buck–Boost Converter with Positive Output Voltage. *IEEE Trans. Ind. Electron.* **2016**, *63*, 2965–2975. [CrossRef]

MDPI

Article

Smart Core and Surface Temperature Estimation Techniques for Health-Conscious Lithium-Ion Battery Management Systems: A Model-to-Model Comparison

Sumukh Surya [1,*], Akash Samanta [2], Vinicius Marcis [2] and Sheldon Williamson [2]

[1] Robert Bosch Engineering and Business Solutions, Bangalore 560100, India
[2] Department of Electrical, Computer and Software Engineering, Ontario Tech University, Oshawa, ON L1G 0C5, Canada; akash.samanta@ontariotechu.net (A.S.); vinicius.marcis@ontariotechu.net (V.M.); sheldon.williamson@uoit.ca (S.W.)
* Correspondence: sumukhsurya@gmail.com

Abstract: Estimation of core temperature is one of the crucial functionalities of the lithium-ion Battery Management System (BMS) towards providing effective thermal management, fault detection and operational safety. It is impractical to measure the core temperature of each cell using physical sensors, while at the same time implementing a complex core temperature estimation strategy in onboard low-cost BMS is also challenging due to high computational cost and the cost of implementation. Typically, a temperature estimation scheme consists of a heat generation model and a heat transfer model. Several researchers have already proposed ranges of thermal models with different levels of accuracy and complexity. Broadly, there are first-order and second-order heat resistor–capacitor-based thermal models of lithium-ion batteries (LIBs) for core and surface temperature estimation. This paper deals with a detailed comparative study between these two models using extensive laboratory test data and simulation study. The aim was to determine whether it is worth investing towards developing a second-order thermal model instead of a first-order model with respect to prediction accuracy considering the modeling complexity and experiments required. Both the thermal models along with the parameter estimation scheme were modeled and simulated in a MATLAB/Simulink environment. Models were validated using laboratory test data of a cylindrical 18,650 LIB cell. Further, a Kalman filter with appropriate process and measurement noise levels was used to estimate the core temperature in terms of measured surface and ambient temperatures. Results from the first-order model and second-order models were analyzed for comparison purposes.

Keywords: electric vehicles; stationary battery energy storage system; battery automated system; online state estimation; thermal modeling; first-order model; second-order model; Kalman filtering

Citation: Surya, S.; Samanta, A.; Marcis, V.; Williamson, S. Smart Core and Surface Temperature Estimation Techniques for Health-Conscious Lithium-Ion Battery Management Systems: A Model-to-Model Comparison. *Energies* **2022**, *15*, 623. https://doi.org/10.3390/en15020623

Academic Editor: Haifeng Dai

Received: 14 November 2021
Accepted: 4 January 2022
Published: 17 January 2022

Publisher's Note: MDPI stays neutral with regard to jurisdictional claims in published maps and institutional affiliations.

1. Introduction

Lithium-ion batteries (LIBs) have been extensively commercialized as a primary energy storage technology for electric vehicles (EVs), stationary energy storage in the smart grid system and several other consumer electronics. The primary dominating factors of LIBs over other energy storage technologies include high energy density, long lifespan, and declining cost [1–4]. However, from literature and practice, it is noticed that the performance of LIBs as well as the durability and reliability are significantly influenced by the operating temperature. Moreover, excessively high temperatures may cause thermal runaway, leading to fire, smoke and other serious safety hazards to the operators [5–7]. Therefore, the requirement of a battery management system (BMS) has become indispensable for effective thermal management and safety of LIB system, which essentially requires accurate information on the core and surface temperature of each cell [8,9] besides other important states such as state of charge (SOC) [10,11] and state of health (SOH). A few other popular functions of an advanced BMS include cell balancing [12,13], fault detection/diagnosis [14] and some other safety inspection functionalities. Several recent research studies highlighted that the accuracy of estimating cell SOC [15], SOH [16] and remaining storage capacity [17] depends on the accurate estimation of cell temperature as all these states are the function of temperature. Moreover, the Columbic efficiency of a cell is greatly affected by the cell temperature during the charging and discharging period. It is worthwhile to mention that the temperature distribution inside the cell is not uniform, and the core temperature remains higher than the surface temperature during practical application, especially under high charging and discharging current [18]. Typically, the difference between the core and surface temperature varies in the range of 5–10 °C [19,20]; however, under high current loading with rapid load fluctuation, it could be even more. Therefore, accurate information on the core and surface temperature is essential to achieving the effective thermal management of an LIB pack besides fault detection. While most of the existing temperature measurement techniques measure the surface temperature directly using physical sensors [21], the measurement of cells' internal temperature is highly challenging when using a physical sensor. Moreover, any high-capacity LIB pack consists of thousands of single LIB cells; thus, installing physical sensors in each cell is not practically feasible from the viewpoint of incremental cost and manufacturing complexity.

To sum up, accurate information on core temperature undoubtedly serves as the essential basis for the thermal management and safety of LIB apart from SOC and SOH estimation whilst it is difficult to measure the core temperature using physical sensors. Therefore, a precise thermal model is crucial to accurate temperature estimation. Moreover, it should be easy to model and computationally inexpensive in order to be implemented in onboard BMS for online prediction of temperature. Several temperature estimation techniques have been proposed by researchers. Typically, a temperature estimation strategy consists of two models, namely, a heat generation model and a heat transfer model [22]. The heat generation model takes physical measurement signals from a cell, typically voltage current, to estimate the total heat generation during charging and discharging. Then, the heat transfer model takes the estimated total heat quantity as model input to predict the temperature of that cell. Depending on the modeling, it can only estimate the core temperature (single-state) or both the core and surface temperature simultaneously (two-state).

Broadly, heat generation models can be classified into three groups, electrochemical models [23–26], data-driven empirical models [27–29] and equivalent circuit models (ECM) [30–32]. Few other researchers have also grouped the heat generation model from the perspective of heat concentration. According to them, the heat generation model could be a concentrated model (all heat is generated at the core), distributed model (heat generated uniformly over the cell) [33] and heterogeneous model [30,34] (due to temperature and current density gradient inside the cell). On the other hand, the heat transfer model can be grouped into finite element analysis (FEA)-based models [32,35–38], lumped multi-node models [27,39–41] and heat capacitive-resistive models [42]. The lumped multi-node model and heat capacitive-resistive models are typically developed based on the analogy between thermal and electrical phenomena. It can be seen that the electrochemical model can produce a very accurate heat generation value provided all model parameters are carefully tuned. However, the electrochemical models are highly complex and computationally expensive. The accuracy of data-driven empirical models highly depends on the experimentally acquired data. Collecting such high-resolution data is challenging, and with the increase in data volume and the number of feature vectors, computational expenses also increase exponentially. On the other hand, an ECM-based estimation model can be designed suitable for online prediction and real-world application by establishing a balance between the computational cost and prediction accuracy. Therefore, ECM-based battery models are extensively used in practice for estimating heat generation in LIB. Further, as far as the heat transfer model is concerned, the heat resistor–capacitor models are easy to develop and computationally efficient compared to FEA-based methods and lumped-parameter multi-node models. The FEA-based methods are highly accurate; however, they come at the expense of a high computational cost. Resistor–capacitor-based models can be optimally engineered to make a balance between prediction accuracy and computational cost depending on the application requirement. Therefore, heat resistor–capacitor model-based temperature estimation is the prime focus of this present study.

Researchers have proposed different kinds of heat resistor–capacitor models for the accurate and precise internal and surface temperature estimation of LIB. However, the major concern regarding the practical application of any model is its computational cost, the capability of online prediction and suitability for onboard BMS. Detailed studies on the thermal characteristics of different layers inside an LIB cell, modelling complexity and the experimental data requirement have been carried out and are listed in the references section [43–48]. The heat resistor–capacitor models use the analogy between thermal and electrical phenomena, where heat capacity (thermal capacitance) and heat transfer coefficient (thermal resistance) are represented as electrical capacitor and resistor, respectively [43]. So far, a first-order (one thermal energy storage element) and second-order (two thermal energy storage elements)-based thermal models have been reported in the literature for temperature estimation. Second-order models are typically complex and require extensive experiments alongside the knowledge of domain experts during modeling. On the other hand, first-order models are easy to implement, computationally inexpensive and require far fewer experiments. Recently, extensive research effort has been made on second-order thermal models of LIB. However, a comparative study between the first-order and second-order model has not yet been assessed. Therefore, this research study focused on the comparative study to investigate whether it is worth investing in developing and implementing a second-order thermal model for the core temperature estimation of LIB in terms of accuracy, modeling complexity and the experimental requirement and its practicability in onboard BMS. Extensive experiments were conducted for data collection, and the data was further utilized for modeling, validation and comparison purposes. The strategy was to employ an ECM-based heat generation model for both a first-order and second-order thermal model to determine the total heat generation inside the cell. A Kalman filter (KF) was used in both the cases to improve prediction performance. Then, the estimation results were compared with the measured data to assess the modeling accuracy. Finally, the predicted results obtained from the first-order and second-order model were compared for the purpose of model-to-model comparison.

The remaining portion of the article is subdivided into five sections for better readability, representation and understanding of the readers. First-order and second-order thermal modeling of LIB and the respective temperature estimation strategy are presented in Section 2. The experimental setup and model parameter identification are discussed in Section 3. Temperature estimation using the fusion of the first-order thermal model with KF and second-order thermal model with KF is described in Section 4. Section 4 also includes the comparative study between the first-order thermal model and second-order thermal model in terms of prediction accuracy and modeling complexity. Major findings and concluding remarks are drawn in the conclusion in Section 5.

2. Thermal Modeling and Temperature Estimation Strategy

Commercially, LIBs are available in many different form factors such as prismatic cells, pouch cells [49] and cylindrical cells. Among these, cylindrical cells are widely used in large-scale high-power applications. However, the cylindrical cell has worse thermal heat dissipation, and the spiral format leads to a big thermal gradient inside the cell. Therefore, the thermal modeling of a 18,650 cylindrical LIB cell is considered in this study, considering the necessity of effective thermal management of cylindrical LIB. The mathematical analysis and the fusion of KF with these thermal models for core and surface temperature estimation are presented in this section. The aim is to provide a guideline for selecting an appropriate thermal model for online prediction with an optimum computational cost suitable for onboard low-cost BMS. As previously discussed in the introduction section, the temperature estimation model consists of one heat generation model and a heat transfer model, where the heat generation model provides input to the heat transfer model. Therefore, the modeling strategy and mathematical analysis of the ECM-based heat generation model are considered here as well.

2.1. Heat Generation Model

The Electric Circuit Model (ECM) [50]—based thermal estimation model has been reported to estimate the total heat generation inside the LIB cell by several researchers. So far, electrochemical modeling has demonstrated the best performance in capturing the nonlinearities of LIB, while at the same time, it is the most complex to model. Capturing the high degree of nonlinearities higher-order ECM is required; however, the computational cost and modeling complexity increase with the increase in model order. Yet the major advantage of ECM is that a balance between the modeling complexity and model accuracy can be achieved through optimization with the help of the model order reduction technique [51,52]. Therefore, a 1-RC (first-order) ECM is considered here to quantify the total heat generation. The 1-RC ECM of LIB is shown in Figure 1. The basic strategy used by any ECM-based heat generation model is to mathematically accumulate the heat generation from internal power losses that typically depend on the internal resistance and charging–discharging current level. Again, the heat generation depends on the cell SOC, current level and temperature, as the internal resistances are the functions of these variables.

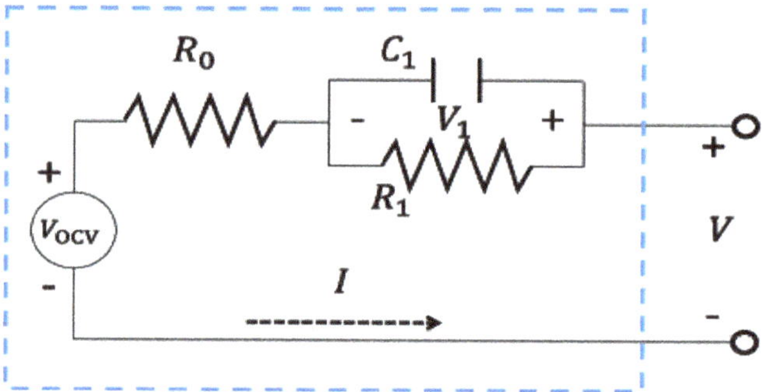

Figure 1. 1-RC ECM (Thevenin's equivalent) model of an LIB cell.

The V_{OCV} and V in Figure 1 represent the open-circuit voltage and the terminal voltage, respectively. The steady-state DC series resistance, which represents the electrolyte resistance to the lithium-ion transportation, is denoted as R_0 in Figure 1. Further, the short transient response is caused by the lithium-ion flow in the solid electrolyte interphase layer, and the anode electrode is represented by polarization resistance (R_1) and capacitance (C_1), respectively. These R_1 and C_1 appear only during the transient period [53]. A 1-RC battery model was considered in this study due to its optimum performance, ease of modelling, low computational cost and adequate accuracy when compared to other higher-order RC models [54,55]. Further, the online determination of heat generation inside LIB with these higher-order models is challenging due to computational cost. For this, Bernardi et al. [56] developed a simplified equation for LIB heat generation calculationthatis suitable for online prediction over other computationally expensive methods such as constant heat generation rate [57], curve fitting technique [58] and Joule's Law [59]-based methods. The governing equation for the total heat generated inside the battery (Q) as developed by Bernardi et al. [56] is shown in Equation (1).

$$Q = I(V - V_{OCV}) \tag{1}$$

The parameters of this equation are also the function of charging–discharging current (I), SOC and temperature, which are estimated using the ECM of the cell. Finally, the value of the Q, obtained from Equation (1), is used as one of the inputs to the first-order and second-order thermal model for temperature estimation, which is discussed in the following section.

2.2. First-Order Thermal Modelling

2.2.1. Mathematical Analysis of First-Order Thermal Model

Now, for the first-order model, as is noted by several other researchers, the surface temperature is considered constant throughout the surface of the cell. Heat transport is only along the radial direction, meaning the lateral surface temperature is considered the same as axial direction (cell temperature at two terminals), as reported in [43]. Further, regarding heat transfer, only heat conduction from the core to the surface is considered. Heat exchange between surface and ambient by convection is not considered. The first-order thermal model is depicted in Figure 2.

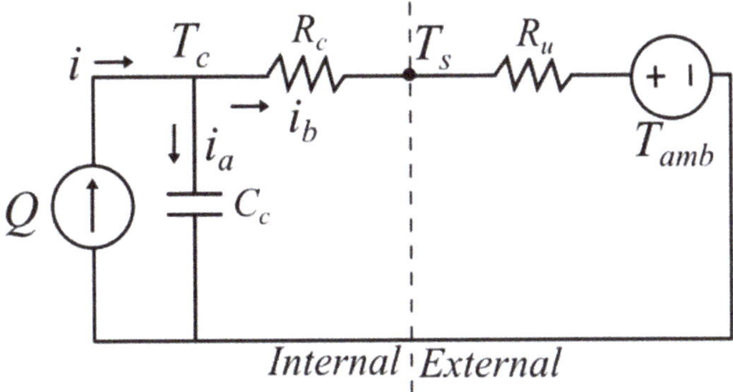

Figure 2. First-order heat resistor–capacitor-based thermal model of LIB.

In Figure 2, the thermal parameters, that is, the heat capacity of the core, heat transfer resistance inside the cell, heat transfer resistance outside the cell and total quantity of heat liberated concentrated from the core, are represented by C_c (J/K), R_c (K/W), R_u (K/W) and Q (J), respectively. The unit of each respective quantity is mentioned in the parentheses. The temperature of the core, surface and ambient is represented by T_c, T_s and T_{amb}, respectively, measured in K. The core temperature at node T_c and surface temperature at node T_s can be monitored using this model; thus this type of model is also referred to as a two-node or two-state thermal model [22,60].The heat resistor–capacitor model uses the analogy between the thermal and electrical systems, as discussed in the introduction section. Thus, for mathematical analysis, the heat transfer rate is represented by electrical current (i), and the branch currents are represented by i_a, i_b in the respective branch, as shown in Figure 2. Therefore, the governing equation of the model can be derived by applying Kirchhoff's Current Law (KCL) at the T_c node. The current balance equation at node T_c reads:

$$i = i_a + i_b = Q \tag{2}$$

Now, by rewriting Equation (2) in terms of thermal parameters, Equation (3) can be found:

$$Q = C_c\frac{dT_c}{dt} + \frac{T_s - T_c}{R_u} + \frac{T_{amb} - T_s}{R_c} \tag{3}$$

By re-arranging Equation (3) we find:

$$C_c\frac{dT_c}{dt} = Q + \frac{T_s - T_c}{R_u} + \frac{T_{amb} - T_s}{R_c} \tag{4}$$

Finally, the value of T_c can be calculated by integrating Equation (4) with respect to the total heat transfer time while the values of T_s and T_{amb} are known. While T_{amb} can be easily measured by employing only one temperature sensor, the measurement of T_s with physical sensors in a high-power LIB pack is challenging. Therefore, the alternative solution is to estimate the surface temperature using a temperature estimation scheme. One such estimation scheme is also proposed in reference [8], which estimates T_s from known T_c.

2.2.2. KF for First-Order Thermal Model

KF is used to estimate and predict an unknown parameter from known parameters. The state model for a KF and the first-order model, as developed in the reference [43,61] and in [62] respectively, are also considered for this study. Now, assuming the state as $T_{c,t}$, output as $T_{s,t}$ and inputs as Q and T_{amb}, The state-space matrices are derived by linearizing Equation (4) in the discrete domain. A linearized version of Equation (4) is shown in Equation (5).

$$T_{c,t} - T_{c,t-1} = \frac{Q_{t-1}}{Cc} + \frac{T_{s,t-1} - T_{c,t-1}}{C_c R_c} + \frac{T_{amb,t} - T_{s,t-1}}{C_c R_u} \quad (5)$$

As shown in reference [8], small changes in T_s can be ignored. Hence, the term $T_{s,t-1}$ can be considered as zero.

$$T_{c,t} = \frac{Q_{t-1}}{Cc} + T_{c,t-1}(1 - \frac{1}{C_c R_c}) + \frac{T_{amb,t-1}}{R_u C_c} \quad (6)$$

The transfer matrices of the KF-based temperature estimation model can be found by reducing Equation (6) in the form of state models as shown in Equations (7)–(9).

Hence,

$$\mathrm{A} = [1 - \frac{1}{C_c R_c}] \quad (7)$$

$$\mathrm{B} = [\frac{1}{C_c} \ \frac{1}{C_c R_u}] \quad (8)$$

$$\mathrm{C} = D = 0 \quad (9)$$

2.3. *Second-Order Thermal Modelling*

2.3.1. Mathematical Analysis of Second-Order Thermal Model

The condition of non-uniform T_s and heat transport in the radial direction through conduction from the core to surface is also considered during the second-order thermal modeling. Additionally, the heat exchange between the surface and ambient is considered in the second-order model, which was not included in the first-order model. Only convective heat exchange between the cell surface and ambient is considered here. Therefore, in addition to the thermal properties of the first-order model, the thermal capacitance of cell case (C_s) is also considered. The resulting equivalent circuit of the second-order thermal model using heat resistor–capacitor is shown in Figure 3, similarly to the findings of other studies [8,60,63,64].

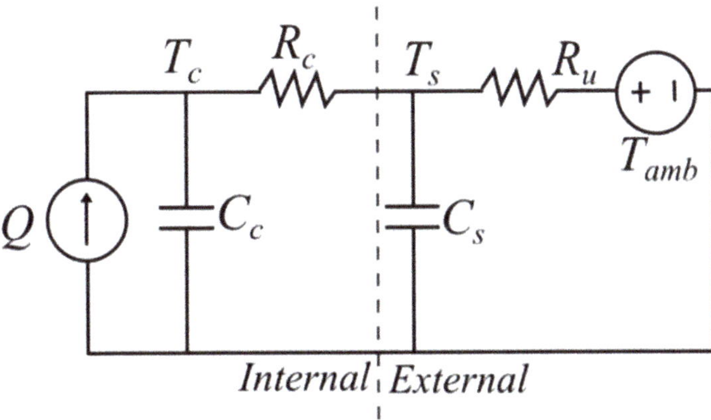

Figure 3. Second-order equivalent circuit thermal model of LIB.

Q needs to be estimated for the same ECM-based strategy mentioned in Section 2.1. To derive the mathematical analysis of the second-order thermal model, heat balance analysis in the core and surface is performed. The heat balance equation at the core and surface is represented in Equations (10) and Equation (11), respectively [8].

$$C_c \frac{dT_c}{dt} = Q + \frac{T_s - T_c}{R_c} \tag{10}$$

$$C_s \frac{dT_s}{dt} = \frac{(T_{amb} - T_s)}{R_u} - \frac{(T_s - T_c)}{R_c} \tag{11}$$

2.3.2. KF for Second-Order Thermal Model

T_c could be estimated by re-arranging the coupled ordinary differential equations of the second-order thermal model. Since the thermal model has two thermal energy storage parameters (C_c and C_s), two governing equations are used to estimate T_c in terms of measured T_s and T_{amb}.

$$\mathbf{A} = [1 - \frac{1}{C_c(R_c + R_u)}] \tag{12}$$

$$\mathbf{B} = [\frac{1}{C_c} \quad \frac{1}{C_c(R_c + R_u)}] \tag{13}$$

$$\mathbf{C} = [\frac{R_u}{R_c + R_u}] \tag{14}$$

$$\mathbf{D} = [0 \quad \frac{R_c}{R_c + R_u}] \tag{15}$$

It is worth noting that C_c, R_c and R_u in the second-order thermal model are the same as C_p, R_{in} and R_{out}, respectively, in the first-order model.

2.3.3. Fundamentals of KF

It is worth providing a basic explanation of KF as it is the heart of the temperature estimation scheme discussed here. A KF is a linear quadratic estimator and is mainly used in statistics and control engineering. It outputs the estimates of an unknown state and uses the noise and the inaccuracies of the measured output. Some of the common examples of KF usage include guidance, navigation and core temperature estimation in EVs; the general form of KF is shown below:

$$X_k = A_{k-1}X_{k-1} + B_{k-1}U_{k-1} + W_{k-1} \tag{16}$$

$$Y_k = C_k X_k + D_k U_k + V_k \tag{17}$$

where X_k is the state of the system ($T_{c,t}$), Y_k is the output of the system ($T_{s,t}$), U_k is the input to the system ($[T_{amb,t}\ Q]^T$), t presents the state of the system and $t-1$ represents the previous state of the system. The block diagram of a KF is shown in Figure 4. It is a robust and simple technique used to estimate data based on its input signal. It uses mathematical modeling of the system and by giving the same input as an actual system, it predicts the output. The measured output from the actual system and predicted output from the mathematical model are then compared to obtain the error. This error is multiplied with Kalman gain and is added to the predicted state to obtain an accurate estimated state [65].

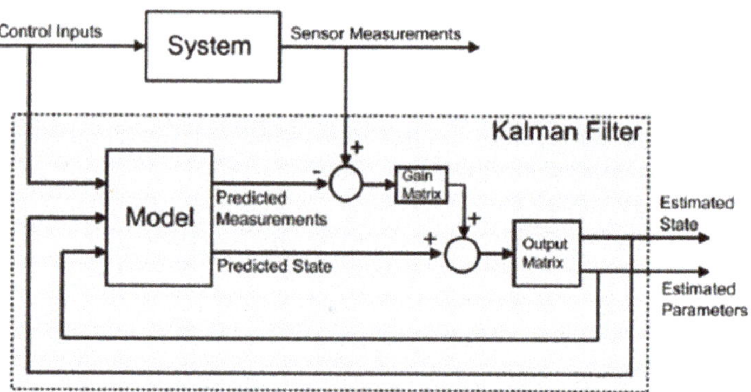

Figure 4. Fundamental building blocks of KF based estimation scheme.

3. Experimental Analysis for Thermal Model Parameterization

An automated battery testing system is the best option to collect battery test data, especially for an LIB, as LIB cells are highly sensitive to voltage, current, temperature and other environmental uncertainties. Therefore, an in-house "Battery Automated System (BAS)" was previously developed by the research group of Smart Transportation Electrification and Energy Research (STEER). The setup was used to invent the constant temperature constant current (CT-CV) charging technique [66,67] and several other prominent research studies in the BMS domain [2,4,8]. The BAS is an experimental setup with a fully programmable test environment control and data acquisition system. A schematic layout of the BAS is shown in Figure 5. The experimental data were then used for the parameter estimation of ECM and thermal modeling, model validation and model-to-model comparison purposes. Interested readers are invited to refer to these papers [2,4,8,66,67] for more details about the experimental setup. However, a brief overview of the experimental setup and test conditions is also mentioned in this section as a quick reference for the readers. The basic idea was to identify the input parameters of the thermal model, that is, heat capacity and heat transfer coefficients, through a steady-state analysis as well as transient experiments based on the nonlinear least square algorithm. The LIB cell was tested at three different temperatures where the internal battery temperature was raised using standard current pulses that were within the permissible limit specified on the manufacturer datasheet to ensure no capacity fade occurred during testing.

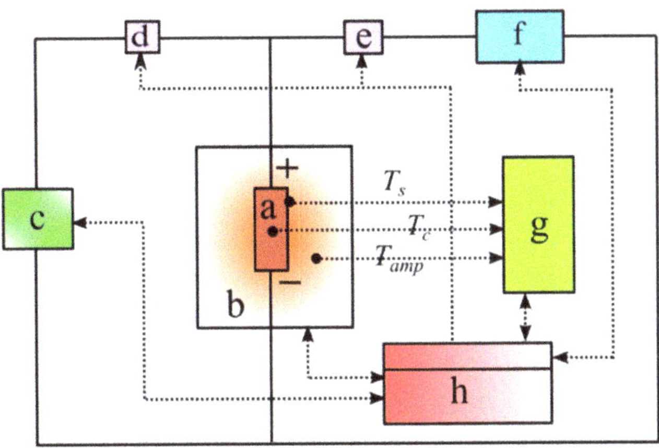

Figure 5. Schematic layout of the Battery Automated System (BAS).

Experimental Setup

Battery testing was performed on a 18,650 NMC (Lithium Nickel Manganese Cobalt Oxide) LIB, manufactured by LG Chem. Detailed specifications of the cell as provided by the manufacturer are mentioned in Table 1. A programmable power supply (Model: E36313A) from Keysight and a programmable electronic load (Model: BK8601) from B&K Precision were used for charging and discharging the battery with a predefined charging–discharging current profile. Further, a programmable temperature chamber was used to maintain the T_{amb} based on a predefined set-point. Finally, to control the BAS a MATLAB script-based program was used. A programmable data acquisition system (DAQ) (model DPM66204) from Chroma was used to collect the cell voltage, current and temperature data. Different current profile sat three different ambient temperatures (T_{amb} = 273 K (0 °C), 293 K (20 °C) and 323 K (50 °C)) were used for charging and discharging experiments. Finally, a nonlinear least square algorithm was used for online parameter estimation for developing the ECM and thermal model as demonstrated by Surya et al. [8]. All the model components were designed in MATLAB using three-dimension interpolated look up tables where the feature vectors were SOC, I_{bat} and T_{amb}. The heat generation model and the first-order and second-order thermal model were also developed in the MATLAB/Simulink and Simscape environment. Finally, an extensive simulation study was conducted to collect the simulated core and surface temperature data for further analysis. Simulation results were used for model validation as well as model-to-model comparison between the first-order and second-order thermal models. The core temperature (T_c) was estimated using a KF for various patterns of currents that were within the permissible limit specified on the manufacturer datasheet to ensure no capacity fade occurred during testing.

Table 1. Specifications of 18,650 LIB cell under test.

Specification Name	Values
Manufacturer and Model	LG Chem/INR18,650HG2
Cell Form Factor	Cylindrical (18,650)
Chemistry	Lithium Nickel Manganese Cobalt Oxide ($LiNiMnCoO_2$)
Nominal Voltage	3.6 V
Nominal Capacity	3 Ah
Standard Charging (CC-CV)	1.5 A, 4.2 V, Cut-off: 50 mA
Fast Charging (CC-CV)	4.0 A, 4.2 V, Cut-off: 100 mA
Discharging Condition	20 A (Max. Current), 2.5 V (Cut-off Voltage)
Operating Temperature	Charge: 0 to 50 °C, Discharge: −30 to 60 °C
Pack Weight	48 g

4. Results and Discussion

This research study intended to answer whether it is worth developing a second-order model instead of a first-order model for online temperature prediction by low-cost onboard BMS, firstly, by developing a first-order and second-order thermal model utilizing battery test data and MATLAB-based online parameter estimation; secondly, by simulating the temperature profile of the cell using the first-order and second-order thermal models subjected to different current profiles. The intention was to investigate the impact of charging–discharging current on the core and surface temperature of the cell. Thirdly, we compared the estimation results obtained from the first-order and second-order models. All simulations were carried out in the MATLAB Simulink environment, where a fixed solver and an appropriate step time were used [62]. Initially, simulation was carried out without employing a KF to deduce the baseline analysis. Figure 6 shows the current profile used for the base case analysis, and Figure 7 shows the plots of estimated T_c, T_s and the measure T_{amb}. Previously, we measured T_s from experiments. By comparing the measured and estimated T_s it was observed that both T_s were within the acceptable limit, and T_c and T_s closely followed the current profile, and $T_c > T_s > T_{amb}$, as per the expectation, confirming the modeling accuracy.

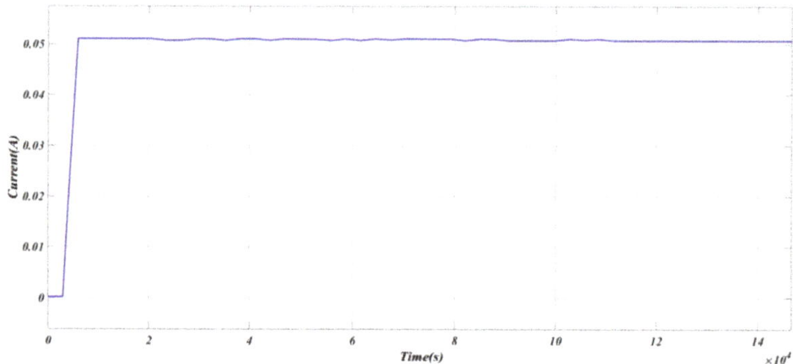

Figure 6. The pattern of the discharging current applied to the cell.

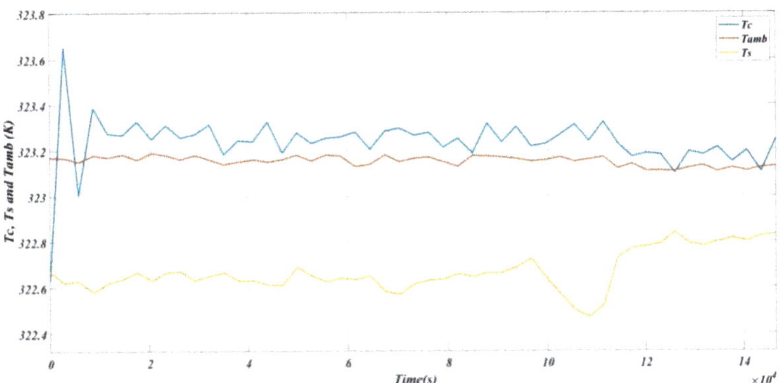

Figure 7. The plot of T_c, T_s and T_{amb} without using KF.

In the subsequent sections, firstly, T_c was estimated using the combined first-order thermal model and KF for three different current profiles and ambient temperatures, which are illustrated in Case 1, Case 2 and Case 3, respectively. Secondly, a similar study was also conducted for the second-order model and finally, the results were compared. All experiments were carried out with different current profiles as per the manufacturer's recommendation to ensure no battery health degradation [58]. In all cases, the initial currents were kept high for rapid charging of the cell.

4.1. Case 1: T_{amb} = 293 K (20 °C)

At first, T_c was initialized to T_s in the simulation as initially, the cell was in a thermal equilibrium state. T_{amb} was considered as 293 K (20 °C), and a very low value of discharging current was applied for the core and surface temperature to rise. The pattern used in Case 1 is shown in Figure 8, and the plot of estimated T_c and measured T_s are shown in Figure 9 whereas the difference between the estimated T_c and measured T_s is shown in Figure 10. It was observed that T_c and T_s closely followed the current pattern, and the maximum difference between estimated T_c and measured T_s was noted as 6.8 K, whilst it was also noticed that for the entire duration, $T_c > T_s$, and the maximum difference occurred when the current was at its peak.

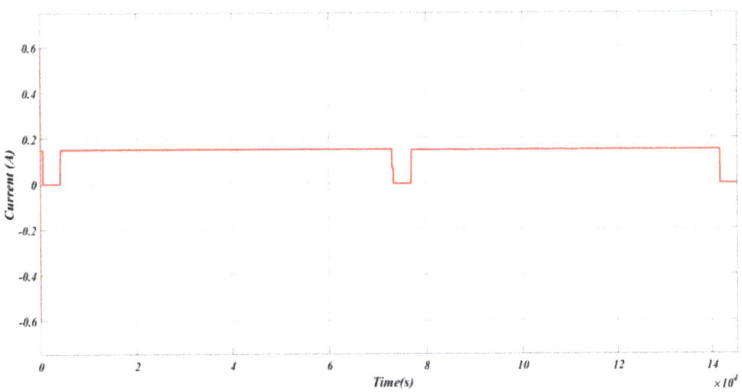

Figure 8. The pattern of the discharging current applied to the cell.

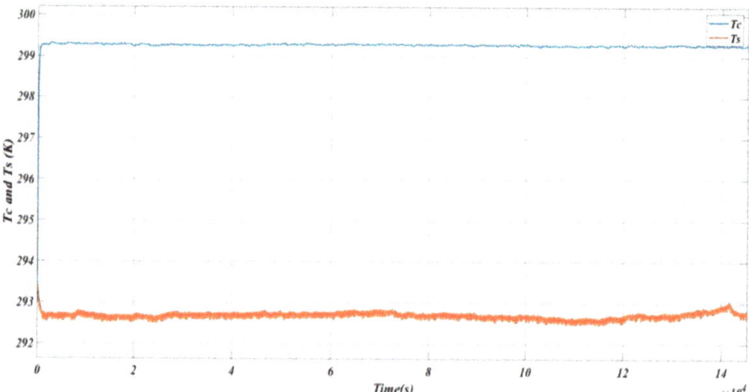

Figure 9. The plot of the estimated T_c and measured T_s.

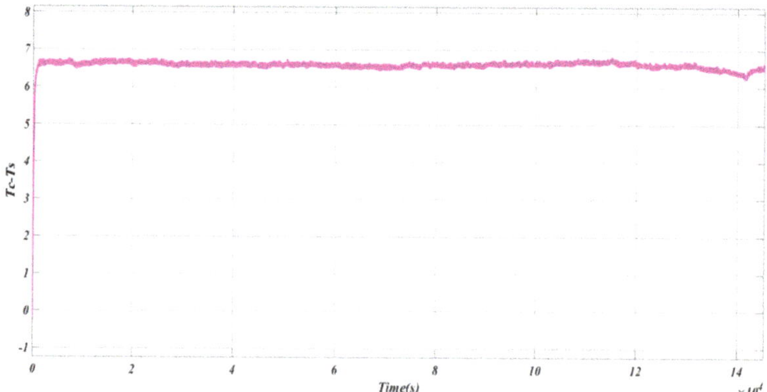

Figure 10. Variation of the difference between the estimated T_c and measured T_s.

4.2. Case 2: T_{amb} = 323 K (50 °C)

In the second phase of the experiments, the temperature of the thermal chamber (T_{amb}) was set to 323 K (50 °C). The pattern used in Case 2 is shown in Figure 11. Similar to Case 1, T_c was initialized to T_s during the simulation here as well. The estimated T_c and measured T_s are shown in Figure 12. It was observed that the temperatures closely followed the current pattern here also. The maximum difference between T_c and measured T_s was noted as 7K. The plot of the difference between the estimated T_c and measured T_s is shown in Figure 13. Similar observations to those made for Case 1 were also noticed here in Case 2 regarding T_c and T_s.

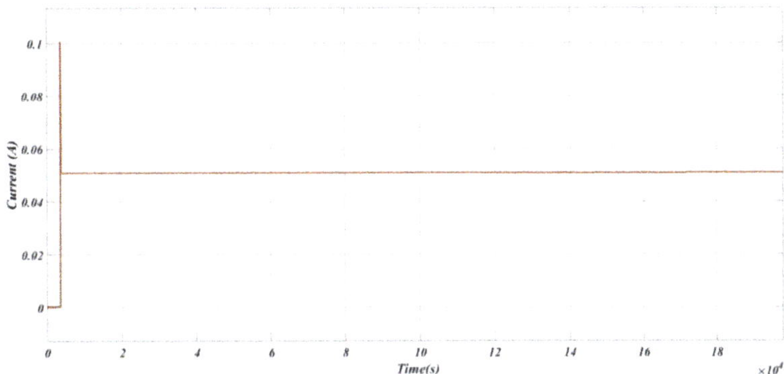

Figure 11. The pattern of the discharging current applied to the cell.

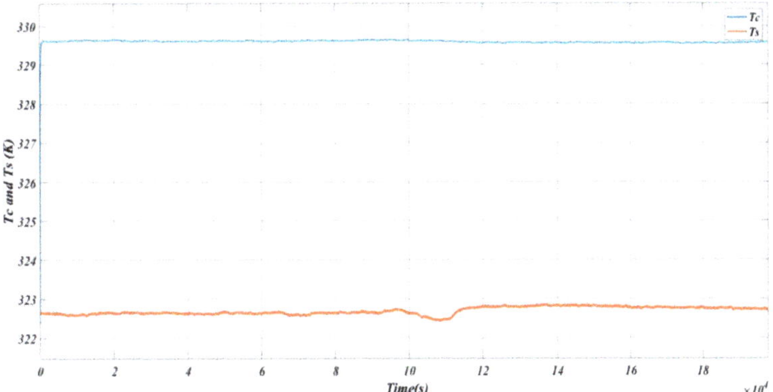

Figure 12. The plot of the estimated T_c and measured T_s.

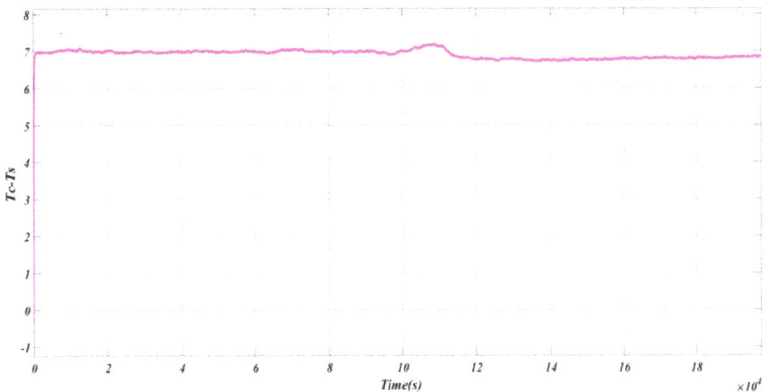

Figure 13. Variation of the difference between the estimated T_c and measured T_s.

4.3. Case 3: T_{amb} =273 K (0 °C)

During Case 3, the temperature of the thermal chamber (T_{amb}) was set to 273 K (0 °C), and T_c was set equal to T_s. The pattern of discharging current applied to the battery is shown in Figure 14. Figure 15 shows the estimated T_c and measured T_s. Figure 16 shows the difference between the estimated T_c and measured T_s. It can be noticed from Figure 15 that at the beginning the magnitude of T_c and T_s were very large. This was due to the high value of discharging current during this period. It was also observed that the temperature rise is a slow process due to the presence of thermal resistances (R_u and R_c). The temperature difference increased as the value of discharge current increased. Therefore, it can be inferred from these observations that the temperature rise closely follows the current through the battery, and the rate of rising of T_c was the same as T_s for a low value of current. However, for higher values of the current the rise in T_c was much higher than that in T_s. From these observations, the importance of accurate core and surface temperature estimation alongside the requirement of effective and efficient thermal management to maintain T_c under the safe operating limit is evidenced.

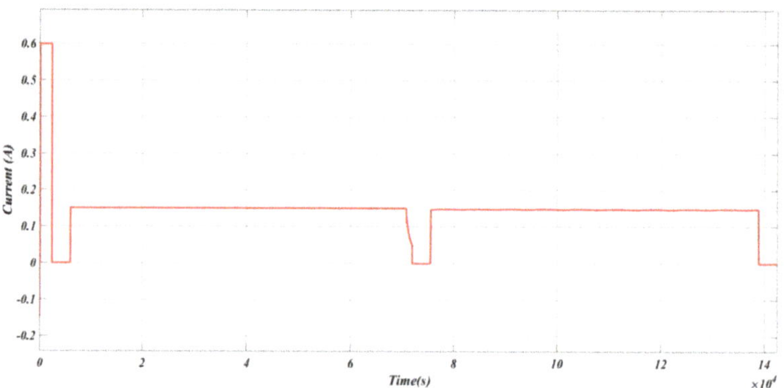

Figure 14. The pattern of the discharging current applied to the cell.

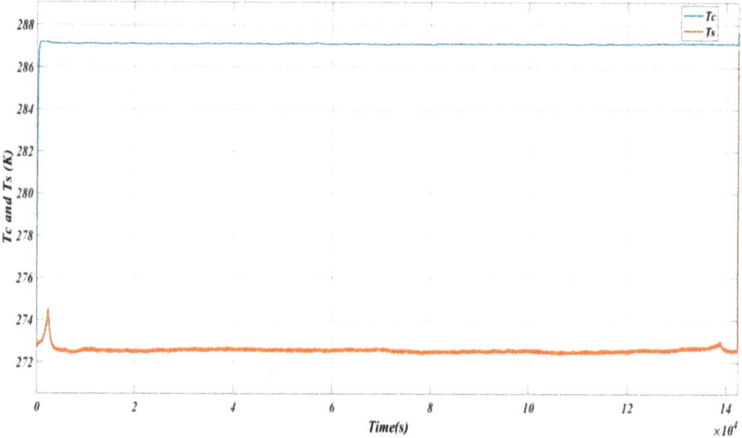

Figure 15. The plot of the estimated T_c and measured T_s.

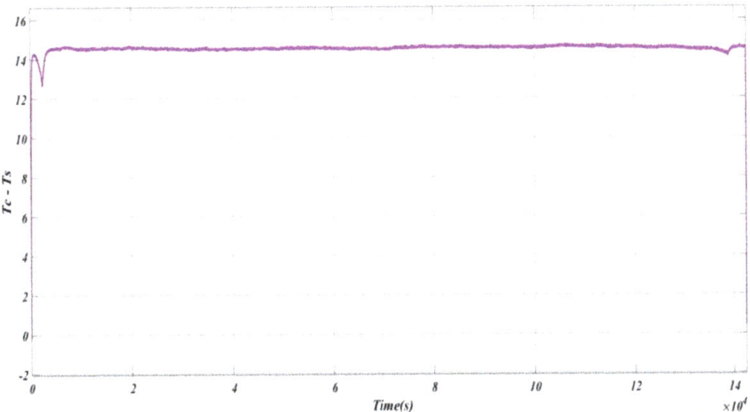

Figure 16. Variation of the difference between the estimated T_c and measured T_s.

4.4. Comparison between First-Order and Second-Order Thermal Models

This section deals with the comparative analysis between the first-order and second-order thermal based on the estimation accuracy, parameter identification, experimental test requirement and suitability for onboard low-cost BMS. To compare the models, the same values of thermal parameters, current, T_s, T_{amb} and Q were injected into the thermal models. Similar current profiles to those used in Case 1, Case 2 and Case 3 of the first-order model were also applied to the second-order thermal model. Heat generation was calculated using the same 1-RC ECM as used in the first-order model. Finally, the estimated T_c profiles obtained from the first-order and second-order thermal models were compared to analyze the prediction accuracy of these models. Figure 17 shows the current profile used for the comparative study, whereas Figures 18 and 19 depict the difference in T_c and T_s obtained from the first-order and second-order thermal model, respectively.

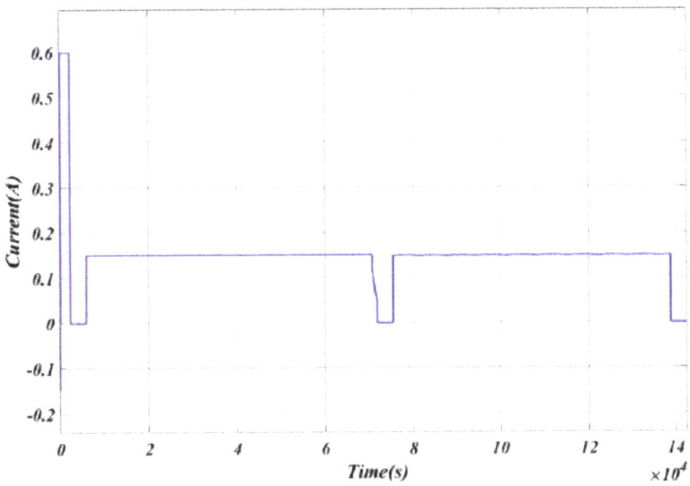

Figure 17. The pattern of the discharging current applied to both the models.

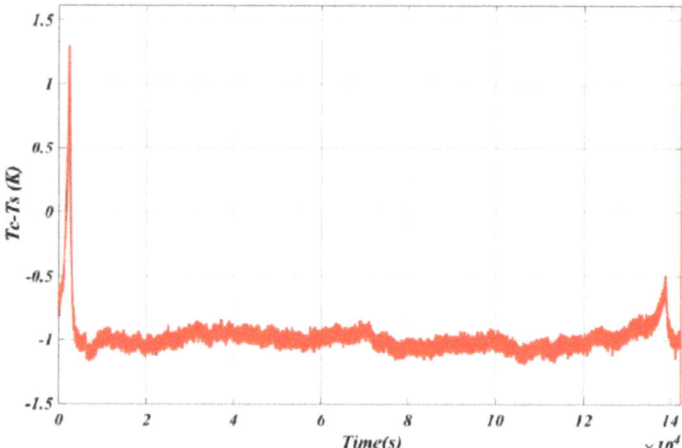

Figure 18. Difference between T_c and T_s obtained from the second-order thermal model.

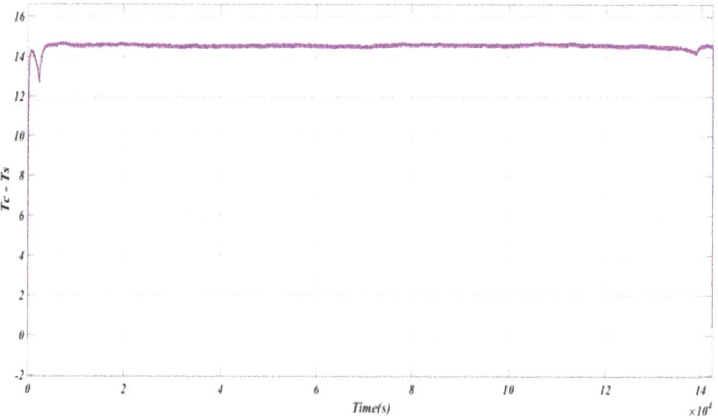

Figure 19. Difference between T_c and T_s obtained from the first-order thermal model.

It was observed that the difference in temperatures was larger in the first-order thermal model due to the change in T_c and not T_s. This is because of the decoupling between T_s and T_c, as seen in Equation (3). Moreover, while comparing Equations (7)–(9) with Equations (12)–(14), it was noticed that the output parameter T_s in KF showed no dependence on the state $T_{c,t-1}$ which is also a major reason behind the estimation error in case of the first-order model. Further, references [43,61] demonstrated that C_c and R_u of the second-order thermal model have a significant effect on T_c. Since these parameters were not present in the C and D matrices of the first-order model, a large increase in T_c was observed. The thermal parameter sensitivity analysis, as conducted in references [8,61], also confirmed the same reason behind the difference in temperature estimation by the first-order thermal model. It was found that the difference between T_s and T_c is increased if the discharge current increases. Hence, for currents with dynamic changes, T_c estimation using the first-order model provides a large difference from the second-order model. Further, C_c only contributed to the transient part of T_c. However, with small changes in R_c and R_u, a large variation in T_c was also observed. The modeling complexity, experimental requirement and computational expenses in the used second-order model were not considerably high compared to the first-order model considered here. A tradeoff between the modeling complexity and accuracy requirement suggests the implementation of a second-order model is worthwhile for smart BMS, especially for high-power applications of LIB.

4.5. Comparison between First-Order and Second-Order Thermal Models for Higher C Rates

As discussed in the introduction, the performance of different types of battery models is highly influenced by the value of charging–discharging current. As was already witnessed from the above discussion, the second-order model is more accurate compared to the first-order model. However, it is equally important to assess the performance of the second-order model in a high value of discharge current for almost all practical purposes a high value of discharge current is used. Therefore, a discharge current of 5A was applied to both the first and second-order thermal models to observe the change in T_c. and T_s. The difference between the estimated T_c and estimated T_s for the first and second-order thermal models is shown in Figure 20.

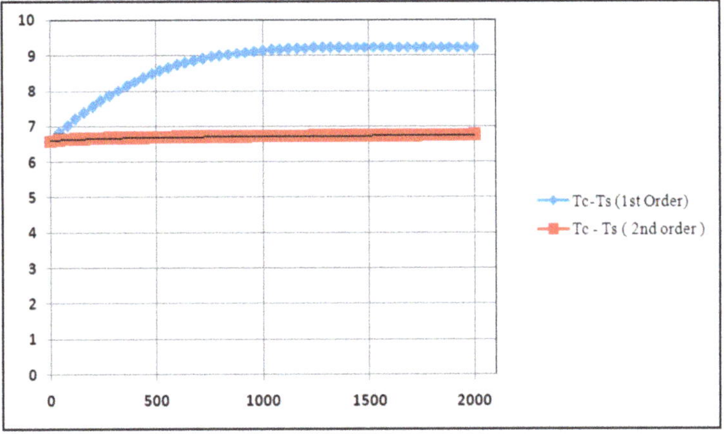

Figure 20. Comparison between T_c-T_s for higher C discharge.

It was observed that the error (T_c-T_s) was higher in the first-order model than in the second-order model. Therefore, it could be concluded that the second-order model can also predict a highly accurate temperature state in practical applications as well.

5. Conclusions

This paper deals with the core temperature (T_c) estimation of lithium-ion 18,650 cell using a Kalman filter (KF). This estimation provides effective thermal management, state estimations, operational safety and the longer useful life of LIB. Initially, a detailed discussion regarding the importance of core and surface temperature estimation was presented followed by a review of the state-of-the-art temperature estimation strategies and thermal modeling of LIB. Equivalent Circuit Models (ECM) of LIB-based heat generation model and heat resistor–capacitor-based thermal models were developed in a MATLAB/Simulink environment. Regarding heat resistor–capacitor-based thermal modeling, one first-order and one second-order thermal model were developed and validated using laboratory experimental data. Further, extensive simulation studies were conducted to demonstrate the influence of battery current and ambient temperature on the core and surface temperature of the LIB cell. The heat transfer equations for a first-order and second-order thermal model were derived, modeled and simulated. KF with appropriate process and measurement noise levels was also used to estimate T_c in terms of measured surface (T_s) and ambient temperature (T_{amb}). Finally, these results were compared to assess the prediction accuracy of these models. The difference between the core and surface temperatures was noted as approximately 7 K to 8 Kin the first-order model, whereas it was only about 1 K to 2 Kin the second-order thermal model. T_s showed no dependence on T_c in the first-order thermal model. Further, the output parameter T_s in KF showed no dependence on the state $T_{c,t-1}$, which is also a major reason behind the estimation error in the case of the first-order model. The thermal capacitance of core (C_c) and resistances (R_u) of the second-order thermal model have a significant effect on T_c. Since these parameters are not present in the C and D matrices of the first-order model, a large increase in T_c was observed in the first-order thermal model. Hence, the inaccuracy was only due to the error in T_c estimation. The findings are also supported by several other research studies in the domain. Further, the consideration of the thermal capacitance of cell casing and the impact of ambient conditions on the second-order model were the reasons for high accuracy. Further, the performances of first and second-order thermals were also judged with a high value of discharge current for assessing their performance during practical operation. It was observed that the second-order model performance was highly satisfactory compared to the first-order model even in practical applications typically requiring a high value of discharge current. However, estimating the additional parameters of the second-order model requires more experimental data and time. Moreover, due to the complex mathematical form of the second-order model, it takes more computation time. However, looking at the prediction accuracy and the increasing stringent requirement of highly accurate states of battery, it could be stated that it is worth investing more time, cost and expertise in developing a second-order thermal model for more accurate temperature estimation in LIB. This is especially true for the advanced BMS required for high-power LIB packs used in EVs and grid-tied energy storage alongside highly sophisticated consumer electronics. The discussed second-order thermal of a single cell can be extended to an LIB pack by integrating the thermal gradient and the impact of peripheral cells alongside optimal placing of temperature sensors inside the battery casing to adjust the ambient temperature parameter value in the model. All these aspects will be considered in our future research.

Author Contributions: Conceptualization, S.S. and S.W.; methodology, S.S. and A.S.; software, S.S. and A.S.; validation, S.S.; formal analysis, S.S. and A.S.; investigation, S.S. and A.S.; resources, V.M. and S.W.; data curation, V.M.; writing—original draft preparation, S.S. and A.S.; writing—review and editing, A.S. and S.W.; visualization, S.S. and A.S.; supervision, S.W.; All authors have read and agreed to the published version of the manuscript.

Funding: This research received no external funding.

Institutional Review Board Statement: Not applicable.

Informed Consent Statement: Not applicable.

Data Availability Statement: Not applicable.

Conflicts of Interest: The authors declare no conflict of interest. No data/information from Robert Bosch Engineering and Business Solutions Private Limited (RBEI) were used for this work.

References

1. Kleiner, J.; Stuckenberger, M.; Komsiyska, L.; Endisch, C. Advanced Monitoring and Prediction of the Thermal State of Intelligent Battery Cells in Electric Vehicles by Physics-Based and Data-Driven Modeling. *Batteries* **2021**, *7*, 31. [CrossRef]
2. Marcis, V.A.; Kelkar, A.; Williamson, S.S. Electrical circuit modeling of a 18650 lithium-ion cell for charging protocol testing for transportation electrification applications. In Proceedings of the 2020 IEEE Transportation Electrification Conference & Expo (ITEC), Chicago, IL, USA, 23–26 June 2020.
3. Dao, V.Q.; Dinh, M.-C.; Kim, C.S.; Park, M.; Doh, C.-H.; Bae, J.H.; Lee, M.-K.; Liu, J.; Bai, Z. Design of an Effective State of Charge Estimation Method for a Lithium-Ion Battery Pack Using Extended Kalman Filter and Artificial Neural Network. *Energies* **2021**, *14*, 2634. [CrossRef]
4. Marcis, V.A.; Praneeth, A.V.J.S.; Patnaik, L.; Williamson, S.S. Analysis of CT-CV Charging Technique for Lithium-ion and NCM 18650 Cells over Temperature Range. In Proceedings of the Proceedings of the IEEE International Conference on Industrial Technology, Buenos Aires, Argentina, 26–28 February 2020.

5. Li, W.; Zhu, J.; Xia, Y.; Gorji, M.B.; Wierzbicki, T. Data-Driven Safety Envelope of Lithium-Ion Batteries for Electric Vehicles. *Joule* **2019**, *3*, 2703–2715. [CrossRef]
6. Lisbona, D.; Snee, T. A review of hazards associated with primary lithium and lithium-ion batteries. *Process Saf. Environ. Prot.* **2011**, *89*, 434–442. [CrossRef]
7. Kong, L.; Li, C.; Jiang, J.; Pecht, M.G. Li-ion battery fire hazards and safety strategies. *Energies* **2018**, *11*, 2191. [CrossRef]
8. Surya, S.; Marcis, V.; Williamson, S. Core Temperature Estimation for a Lithium ion 18650 Cell. *Energies* **2020**, *14*, 87. [CrossRef]
9. Zhu, J.; Sun, Z.; Wei, X.; Dai, H. Battery internal temperature estimation for LiFePO4 battery based on impedance phase shift under operating conditions. *Energies* **2017**, *10*, 60. [CrossRef]
10. Sidhu, M.S.; Ronanki, D.; Williamson, S. State of charge estimation of lithium-ion batteries using hybrid machine learning technique. In Proceedings of the IECON Proceedings (Industrial Electronics Conference), Lisbon, Portugal, 14–17 October 2019.
11. Sidhu, M.S.; Ronanki, D.; Williamson, S. Hybrid State of Charge Estimation Approach for Lithium-ion Batteries using k-Nearest Neighbour and Gaussian Filter-based Error Cancellation. *IEEE Int. Symp. Ind. Electron.* **2019**, *2019*, 1506–1511. [CrossRef]
12. Samanta, A.; Chowdhuri, S. Active Cell Balancing of Lithium-ion Battery Pack Using Dual DC-DC Converter and Auxiliary Lead-acid Battery. *J. Energy Storage* **2021**, *33*, 102109. [CrossRef]
13. Kelkar, A.; Dasari, Y.; Williamson, S.S. A Comprehensive Review of Power Electronics Enabled Active Battery Cell Balancing for Smart Energy Management. In Proceedings of the 2020 IEEE International Conference on Power Electronics, Smart Grid and Renewable Energy (PESGRE2020), Cochin, India, 2–4 January 2020.
14. Samanta, A.; Chowdhuri, S.; Williamson, S.S. Machine Learning-Based Data-Driven Fault Detection/Diagnosis of Lithium-Ion Battery: A Critical Review. *Electronics* **2021**, *10*, 1309. [CrossRef]
15. Tanim, T.R.; Rahn, C.D.; Wang, C.Y. State of charge estimation of a lithium ion cell based on a temperature dependent and electrolyte enhanced single particle model. *Energy* **2015**, *80*, 731–739. [CrossRef]
16. Farmann, A.; Sauer, D.U. A study on the dependency of the open-circuit voltage on temperature and actual aging state of lithium-ion batteries. *J. Power Sources* **2017**, *347*, 1–13. [CrossRef]
17. Zheng, F.; Jiang, J.; Sun, B.; Zhang, W.; Pecht, M. Temperature dependent power capability estimation of lithium-ion batteries for hybrid electric vehicles. *Energy* **2016**, *113*, 64–75. [CrossRef]
18. Surya, S.; Mn, A. Effect of Fast Discharge of a Battery on its Core Temperature. In Proceedings of the 2020 International Conference on Futuristic Technologies in Control Systems & Renewable Energy (ICFCR), Malappuram, India, 23–24 September 2020. [CrossRef]
19. Li, Z.; Zhang, J.; Wu, B.; Huang, J.; Nie, Z.; Sun, Y.; An, F.; Wu, N. Examining temporal and spatial variations of internal temperature in large-format laminated battery with embedded thermocouples. *J. Power Sources* **2013**, *241*, 536–553. [CrossRef]
20. Robinson, J.B.; Darr, J.A.; Eastwood, D.S.; Hinds, G.; Lee, P.D.; Shearing, P.R.; Taiwo, O.O.; Brett, D.J.L. Non-uniform temperature distribution in Li-ion batteries during discharge-A combined thermal imaging, X-ray micro-tomography and electrochemical impedance approach. *J. Power Sources* **2014**, *252*, 51–57. [CrossRef]
21. Raijmakers, L.H.J.; Danilov, D.L.; Eichel, R.A.; Notten, P.H.L. A review on various temperature-indication methods for Li-ion batteries. *Appl. Energy* **2019**, *240*, 918–945. [CrossRef]
22. Pan, Y.W.; Hua, Y.; Zhou, S.; He, R.; Zhang, Y.; Yang, S.; Liu, X.; Lian, Y.; Yan, X.; Wu, B. A computational multi-node electro-thermal model for large prismatic lithium-ion batteries. *J. Power Sources* **2020**, *459*, 228070. [CrossRef]
23. Sun, F.; Xiong, R.; He, H.; Li, W.; Aussems, J.E.E. Model-based dynamic multi-parameter method for peak power estimation of lithium-ion batteries. *Appl. Energy* **2012**, *96*, 378–386. [CrossRef]
24. Lu, L.; Han, X.; Li, J.; Hua, J.; Ouyang, M. A review on the key issues for lithium-ion battery management in electric vehicles. *J. Power Sources* **2013**, *226*, 272–288. [CrossRef]
25. Ghalkhani, M.; Bahiraei, F.; Nazri, G.A.; Saif, M. Electrochemical–Thermal Model of Pouch-type Lithium-ion Batteries. *Electrochim. Acta* **2017**, *247*, 569–587. [CrossRef]
26. Yang, X.G.; Leng, Y.; Zhang, G.; Ge, S.; Wang, C.Y. Modeling of lithium plating induced aging of lithium-ion batteries: Transition from linear to nonlinear aging. *J. Power Sources* **2017**, *60*, 28–40. [CrossRef]
27. Allafi, W.; Zhang, C.; Uddin, K.; Worwood, D.; Dinh, T.Q.; Ormeno, P.A.; Li, K.; Marco, J. A lumped thermal model of lithium-ion battery cells considering radiative heat transfer. *Appl. Therm. Eng.* **2018**, *143*, 472–481. [CrossRef]
28. Esmaeili, J.; Jannesari, H. Developing heat source term including heat generation at rest condition for Lithium-ion battery pack by up scaling information from cell scale. *Energy Convers. Manag.* **2017**, *139*, 194–205. [CrossRef]
29. Arora, S.; Shen, W.; Kapoor, A. Neural network based computational model for estimation of heat generation in LiFePO4 pouch cells of different nominal capacities. *Comput. Chem. Eng.* **2017**, *101*, 81–94. [CrossRef]
30. Chen, M.; Bai, F.; Song, W.; Lv, J.; Lin, S.; Feng, Z.; Li, Y.; Ding, Y. A multilayer electro-thermal model of pouch battery during normal discharge and internal short circuit process. *Appl. Therm. Eng.* **2017**, *120*, 506–516. [CrossRef]
31. Zhao, Y.; Diaz, L.B.; Patel, Y.; Zhang, T.; Offer, G.J. How to Cool Lithium Ion Batteries: Optimising Cell Design using a Thermally Coupled Model. *J. Electrochem. Soc.* **2019**, *166*, A2849–A2859. [CrossRef]
32. Damay, N.; Forgez, C.; Bichat, M.P.; Friedrich, G. Thermal modeling of large prismatic LiFePO4/graphite battery. Coupled thermal and heat generation models for characterization and simulation. *J. Power Sources* **2015**, *283*, 37–45. [CrossRef]
33. Hu, X.; Liu, W.; Lin, X.; Xie, Y. A Comparative Study of Control-Oriented Thermal Models for Cylindrical Li-Ion Batteries. *IEEE Trans. Transp. Electrif.* **2019**, *5*, 1237–1253. [CrossRef]

34. Xie, Y.; Li, W.; Hu, X.; Zou, C.; Feng, F.; Tang, X. Novel Mesoscale Electrothermal Modeling for Lithium-Ion Batteries. *IEEE Trans. Power Electron.* **2020**, *35*, 2595–2614. [CrossRef]
35. Friesen, A.; Mönnighoff, X.; Börner, M.; Haetge, J.; Schappacher, F.M.; Winter, M. Influence of temperature on the aging behavior of 18650-type lithium ion cells: A comprehensive approach combining electrochemical characterization and post-mortem analysis. *J. Power Sources* **2017**, *342*, 88–97. [CrossRef]
36. Fan, Y.; Bao, Y.; Ling, C.; Chu, Y.; Tan, X.; Yang, S. Experimental study on the thermal management performance of air cooling for high energy density cylindrical lithium-ion batteries. *Appl. Therm. Eng.* **2019**, *155*, 96–109. [CrossRef]
37. Saw, L.H.; Poon, H.M.; Thiam, H.S.; Cai, Z.; Chong, W.T.; Pambudi, N.A.; King, Y.J. Novel thermal management system using mist cooling for lithium-ion battery packs. *Appl. Energy* **2018**, *223*, 146–158. [CrossRef]
38. Liu, B.; Yin, S.; Xu, J. Integrated computation model of lithium-ion battery subject to nail penetration. *Appl. Energy* **2016**, *183*, 278–289. [CrossRef]
39. Doyle, M. Modeling of Galvanostatic Charge and Discharge of the Lithium/Polymer/Insertion Cell. *J. Electrochem. Soc.* **1993**, *140*, 1526–1533. [CrossRef]
40. Xiao, Y.; Fahimi, B. State-space based multi-nodes thermal model for lithium-ion battery. In Proceedings of the 2014 IEEE Transportation Electrification Conference and Expo: Components, Systems, and Power Electronics-From Technology to Business and Public Policy (ITEC 2014), Dearborn, MI, USA, 15–18 June 2014.
41. Tian, N.; Fang, H.; Wang, Y. 3-D Temperature Field Reconstruction for a Lithium-Ion Battery Pack: A Distributed Kalman Filtering Approach. *IEEE Trans. Control Syst. Technol.* **2019**, *27*, 847–854. [CrossRef]
42. Ruan, H.; Jiang, J.; Sun, B.; Gao, W.; Wang, L.; Zhang, W. Online estimation of thermal parameters based on a reduced wide-temperature-range electro-thermal coupled model for lithium-ion batteries. *J. Power Sources* **2018**, *396*, 715–724. [CrossRef]
43. Forgez, C.; Vinh Do, D.; Friedrich, G.; Morcrette, M.; Delacourt, C. Thermal modeling of a cylindrical LiFePO4/graphite lithium-ion battery. *J. Power Sources* **2010**, *195*, 2961–2968. [CrossRef]
44. Dees, D.W.; Battaglia, V.S.; Bélanger, A. Electrochemical modeling of lithium polymer batteries. *J. Power Sources* **2002**, *110*, 310–320. [CrossRef]
45. Lin, X.; Perez, H.E.; Siegel, J.B.; Stefanopoulou, A.G.; Li, Y.; Anderson, R.D.; Ding, Y.; Castanier, M.P. Online parameterization of lumped thermal dynamics in cylindrical lithium ion batteries for core temperature estimation and health monitoring. *IEEE Trans. Control Syst. Technol.* **2013**, *21*, 1745–1755. [CrossRef]
46. Choi, J.W.; Aurbach, D. Promise and reality of post-lithium-ion batteries with high energy densities. *Nat. Rev. Mater.* **2016**, *1*, 16013. [CrossRef]
47. Yang, F.; Wang, D.; Zhao, Y.; Tsui, K.L.; Bae, S.J. A study of the relationship between coulombic efficiency and capacity degradation of commercial lithium-ion batteries. *Energy* **2018**, *145*, 486–495. [CrossRef]
48. Wang, Q.; Zhao, X.; Ye, J.; Sun, Q.; Ping, P.; Sun, J. Thermal response of lithium-ion battery during charging and discharging under adiabatic conditions. *J. Therm. Anal. Calorim.* **2016**, *124*, 417–428. [CrossRef]
49. An, S.J.; Li, J.; Daniel, C.; Kalnaus, S.; Wood, D.L. Design and Demonstration of Three-Electrode Pouch Cells for Lithium-Ion Batteries. *J. Electrochem. Soc.* **2017**, *164*, A1755–A1764. [CrossRef]
50. Patnaik, L.; Williamson, S. A Five-Parameter Analytical Curvefit Model for Open-Circuit Voltage Variation with State-of-Charge of a Rechargeable Battery. In Proceedings of the 2018 IEEE International Conference on Power Electronics, Drives and Energy Systems (PEDES 2018), Chennai, India, 18–21 December 2018.
51. Tran, N.T.; Farrell, T.; Vilathgamuwa, M.; Choi, S.S.; Li, Y. A Computationally Efficient Coupled Electrochemical-Thermal Model for Large Format Cylindrical Lithium Ion Batteries. *J. Electrochem. Soc.* **2019**, *166*, A3059–A3071. [CrossRef]
52. Subramaniam, A.; Kolluri, S.; Santhanagopalan, S.; Subramanian, V.R. An Efficient Electrochemical-Thermal Tanks-in-Series Model for Lithium-Ion Batteries. *J. Electrochem. Soc.* **2020**, *167*, 113506. [CrossRef]
53. Bryden, T.S.; Dimitrov, B.; Hilton, G.; Ponce de León, C.; Bugryniec, P.; Brown, S.; Cumming, D.; Cruden, A. Methodology to determine the heat capacity of lithium-ion cells. *J. Power Sources* **2018**, *395*, 369–378. [CrossRef]
54. Lin, C.; Yu, Q.; Xiong, R.; Wang, L.Y. A study on the impact of open circuit voltage tests on state of charge estimation for lithium-ion batteries. *Appl. Energy* **2017**, *205*, 892–902. [CrossRef]
55. Surya, S.; Channegowda, J.; Datar, S.D.; Jha, A.S.; Victor, A. Accurate battery modeling based on pulse charging using MATLAB/Simulink. In Proceedings of the 9th IEEE International Conference on Power Electronics, Drives and Energy Systems (PEDES 2020), Jaipur, India, 16–19 December 2020.
56. Bernardi, D.; Pawlikowski, E.; Newman, J. A general energy balance for battery systems. *J. Electrochem. Soc.* **1985**, *132*, 5. [CrossRef]
57. Greco, A.; Cao, D.; Jiang, X.; Yang, H. A theoretical and computational study of lithium-ion battery thermal management for electric vehicles using heat pipes. *J. Power Sources* **2014**, *257*, 344–355. [CrossRef]
58. Rao, Z.; Wang, S.; Zhang, G. Simulation and experiment of thermal energy management with phase change material for ageing LiFePO4 power battery. *Energy Convers. Manag.* **2011**, *52*, 3408–3414. [CrossRef]
59. Xie, Y.; Shi, S.; Tang, J.; Wu, H.; Yu, J. Experimental and analytical study on heat generation characteristics of a lithium-ion power battery. *Int. J. Heat Mass Transf.* **2018**, *122*, 884–894. [CrossRef]
60. Chen, L.; Hu, M.; Cao, K.; Li, S.; Su, Z.; Jin, G.; Fu, C. Core temperature estimation based on electro-thermal model of lithium-ion batteries. *Int. J. Energy Res.* **2020**, *44*, 5320–5333. [CrossRef]

61. Surya, S.; Bhesaniya, A.; Gogate, A.; Ankur, R.; Patil, V. Development of thermal model for estimation of core temperature of batteries. *Int. J. Emerg. Electr. Power Syst.* **2020**, *21*, 20200070. [CrossRef]
62. Surya, S.; Patil, V. Cuk Converter as an Efficient Driver for LED. In Proceedings of the 4th International Conference on Electrical, Electronics, Communication, Computer Technologies and Optimization Techniques (ICEECCOT 2019), Mysuru, India, 13–14 December 2019.
63. Lin, X.; Stefanopoulou, A.G.; Perez, H.E.; Siegel, J.B.; Li, Y.; Anderson, R.D. Quadruple adaptive observer of the core temperature in cylindrical Li-ion batteries and their health monitoring. *Proc. Am. Control Conf.* **2012**, *578*, 578–583. [CrossRef]
64. Sun, J.; Wei, G.; Pei, L.; Lu, R.; Song, K.; Wu, C.; Zhu, C. Online internal temperature estimation for lithium-ion batteries based on Kalman filter. *Energies* **2015**, *8*, 4400–4415. [CrossRef]
65. Plett, G. *BatteryManagement Systems, Volume II: Equivalent-Circuit Methods*; Artech: Morristown, NJ, USA, 2015; ISBN 9781630810283.
66. Patnaik, L.; Praneeth, A.V.J.S.; Williamson, S.S. A Closed-Loop Constant-Temperature Constant-Voltage Charging Technique to Reduce Charge Time of Lithium-Ion Batteries. *IEEE Trans. Ind. Electron.* **2019**, *66*, 1059–1067. [CrossRef]
67. Williamson, S.; Vincent, D.; Praneeth, A.; Sang, P.H. Charging Strategies for Electrified Transport. In *Advances in Carbon Management Technologies*; CRC Press: Boca Raton, FL, USA, 2021; pp. 284–302.

Article

Controlled Energy Flow in Z-Source Inverters

Zbigniew Rymarski, Krzysztof Bernacki * and Łukasz Dyga

Department of Electronics, Electrical Engineering and Microelectronics, Faculty of Automatic Control, Electronics and Computer Science, Silesian University of Technology, Akademicka 16, 44-100 Gliwice, Poland; zbigniew.rymarski@polsl.pl (Z.R.); lukasz.dyga@polsl.pl (Ł.D.)
* Correspondence: krzysztof.bernacki@polsl.pl

Abstract: This paper proposes a method to reduce the output voltage distortions in voltage source inverters (VSI) working with impedance networks. The three main reasons for the voltage distortions include a discontinuous current in the coils of the impedance network, the double output frequency harmonics in the VSI's voltage output caused by insufficient capacitance in the impedance network, and voltage drops on the bridge switches during the shoot-through time. The first of these distortions can be reduced by increasing the current of the impedance network when the output VSI current is low. This method requires storing energy in the battery connected to the DC link of the VSI during the "non-shoot through" time. Furthermore, this solution can also be used when the Z-source inverter works with a photovoltaic cell to help it attain a maximum power point. The Z-source inverter is essentially a voltage source inverter with the Z-source in the input. In this paper, the theory behind basic impedance networks of Z-source and quasi-Z-source (qZ-source) is investigated where simulations of the presented solutions and experimental verification of the results are also presented.

Keywords: impedance network; Z-source; quasi-Z-source; voltage source inverter; voltage distortions

Citation: Rymarski, Z.; Bernacki, K.; Dyga, Ł. Controlled Energy Flow in Z-Source Inverters. *Energies* **2021**, *14*, 7272. https://doi.org/10.3390/en14217272

Academic Editors: Sheldon Williamson and Andrei Blinov

Received: 6 October 2021
Accepted: 1 November 2021
Published: 3 November 2021

1. Introduction

The Z-source impedance network was proposed initially by Peng [1]. This type of DC/DC converter was increasing the input DC voltage that is connected to a single-phase or three-phase bridge voltage source inverter (VSI) which switches were used to store energy in the coils of a Z-source. During shoot-through time, energy is stored when both switches in one of the inverter bridge legs are activated. This is only possible only in zero states of the inverter. The modulation index M is restricted to the equation $M = 1 - d_Z$ where $d_Z = T_{ST}/T_s$. The parameters T_{ST}, T_s, and d_Z represent the shoot-through time, switching period of the inverter, and shoot-through time coefficient, respectively.

For a Z-source, it is essential that the shoot-through time, d_Z is less than 0.5. A voltage source inverter with a Z-source is known as the Z-source inverter (ZSI). An impedance network can function simply as a DC/DC converter with one additional switch in its output realizing shoot-through time but without an inverter. The input current of the Z-source is discontinuous (discontinuous input current—DIC) so Peng showed the changed structure of the impedance network [2,3]. When a diode usually connected in series with the input is replaced, this structure is called a qZ-source. As a result of this modification, the new quasi-Z-source inverter (qZSI) structure is characterized by a continuous input current (CIC) which has improved the use of an impedance network in photovoltaic (PV) systems [4]. Various methods of improving impedance networks structures have been developed [5] and a suitable example is the switched inductor Z-source inverter (SLZSI) [6]. The benefit of using these improved converters is a higher boost factor of the input DC voltage than in the qZSI. Other existing impedance network structures include the embedded SLZSI [7], an inductor-capacitor-capacitor-transformer Z-source (LCCTZSI) [8,9], and a cascaded quasi-Z-source (CqZSI) [10]. The two-winding magnetically coupled impedance source (MCIS) impedance network with a continuous input current [11] has a high boost factor. The impedance network circuit based on three coupled inductors with a delta (Δ) connection is presented in [12] and further developed in [13]. The networks found in references [11] and [12] respectively were functional where an additional switch was used without an inverter. A broad review of the impedance network topologies is presented in [14,15], amongst other newly developed solutions based on impedance networks [16–20]. Additionally, several methods of controlling impedance networks have been considered which can be reviewed in [21,22]. However, the symmetric structure of a Z-source with discontinuous input current due to a diode connected in series (Figure 1), and an asymmetric quasi-Z-source (Figure 2) with maximum boost control is sufficient to show the influence of an impedance network on VSI output voltage distortions and proposed ways of reducing these distortions.

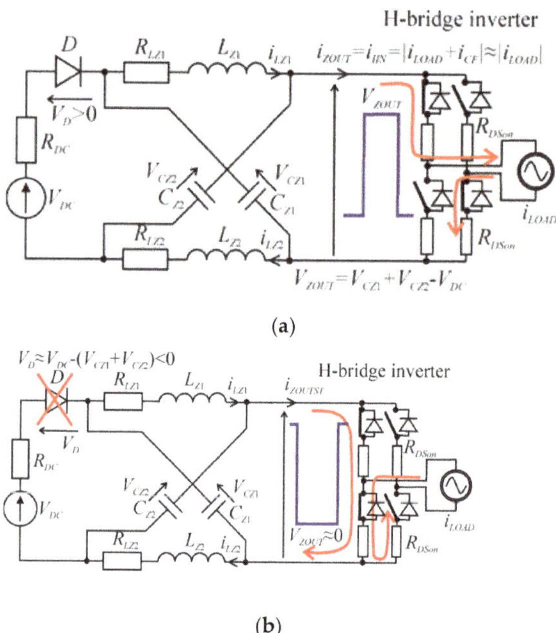

Figure 1. (**a**) Non-shoot-through state and (**b**) shoot-through state of the Z-source impedance network with the VSI.

Further investigation of these improved network structures has shown that the power efficiency of these systems including the decreased efficiency of the inverter is lower than the efficiency of basic structures. Owing to this decreased efficiency the real boost factor is also much lower than expected [23]. It is worth mentioning that significant differences in recorded levels of radiated disturbances can be expected depending on the type of impedance network structure used [24]. Unfortunately, additional losses in the switches of the VSI during the shoot-through time are observed when switches are absent in the impedance networks. Comparing the performance of a boost converter [23,25], it can be shown that the VSI with an input synchronous boost converter can have a higher efficiency than the same inverter with an impedance network.

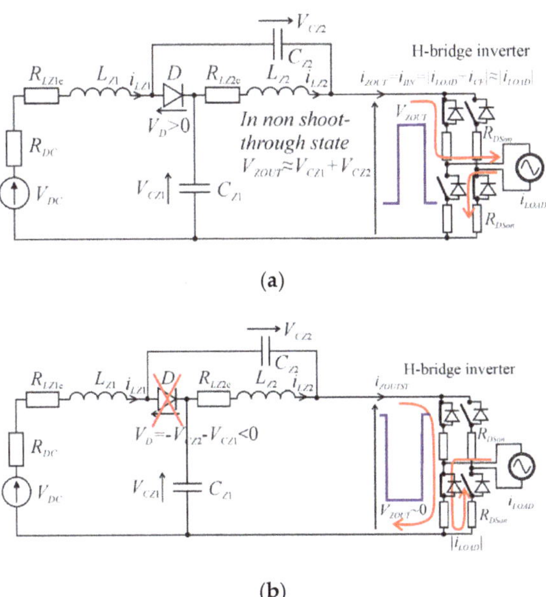

Figure 2. (**a**) Non-shoot-through state and (**b**) shoot-through state of the qZ-source impedance network with the VSI.

The basic structures of Z-source and qZ-source impedance networks are utilized today in photovoltaic systems [26]. The main disadvantage of these impedance networks lies in the discontinuous current mode (DCM) where the current in the inductors is equal to zero for a time period during T_s where there is a low load of the VSI and a low d_Z coefficient. This is the main reason for the VSI output voltage distortions as shown in Figure 3a,b. By calculating a sufficiently large inductance of the coils [23,27,28] and selecting an appropriate magnetic material [29] for the lowest load while assuming the value of d_Z, the current in the coils should not decrease to zero. During operation, it cannot be guaranteed that the load current will be nominal. Thus, the additional current taken from the impedance network is a solution of DCM omitting for a low load current.

Another reason for output distortions is the insufficient capacity of Z-source capacitors. Input current from a VSI bridge is like a "rectified" waveform that is filtered by the LC input network and is approximately the first harmonic of the "rectified" current at 100 Hz. This means that 100 Hz distortion is present in the 50 Hz output waveform as shown in Figure 3c. For the insufficient capacity, the output sinusoidal waveform is left-skewed [23,27]. The third type of VSI output distortions are observed after crossing zero output voltage caused by the additional voltage drops on the switched-on transistors during the shoot-through time (see Figure 3a–c), thus causing oscillations after a change of polarization in the PWM voltage. The impedance network influences the dynamic properties of an entire ZSI [23,27,28] which introduces additional resonant frequencies and the additional damping to the Bode plots of the ZSI. The main objective of this paper is to demonstrate how charging the battery from a DC-link after the impedance network during the non-shoot through times can reduce output distortions caused by the DCM of the impedance network. However, charging a battery with too high a current can lead to distortions of the output voltage after the voltage current is zero crossing and oscillations as the result of the higher voltage drops on the switches during the shoot-through time. Experimental results presented will show how charging the battery for a Z-source decreases the output of total harmonic distortions (THD) even in the case when a sophisticated feedback loop, for example, a passivity-based control (PBC), is used.

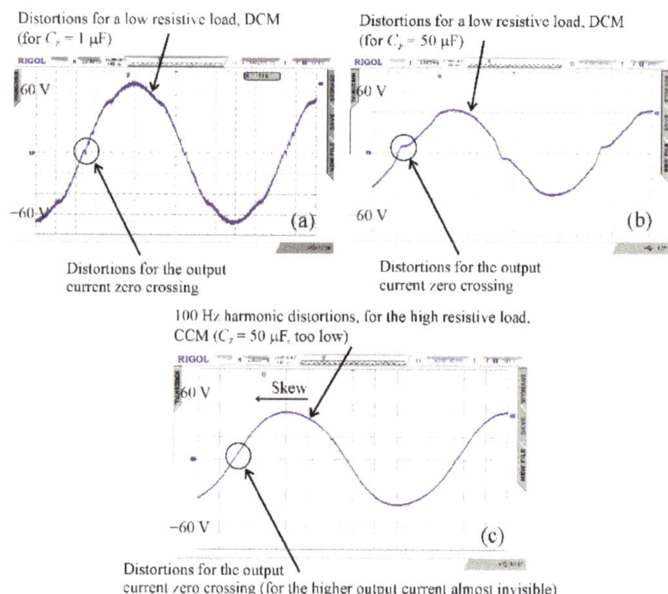

Figure 3. Inverter output voltage distortions, (**a**) Z-source in DCM using a VSI output filter capacitor C_F = 1 μF, (**b**) Z-source in DCM using a VSI output filter capacitor C_F = 50 μF, (**c**) 100 Hz harmonic distortions with a Z-source capacitor C_Z = 100 μF.

Figure 3 presents the different types of output voltage distortions of the ZSI. In Figure 3a,b, the DCM of the Z-source uses a low load current and ZSI output filter capacitors of C_F = 1 μF and 50 μF respectively. Figure 3c shows the distortions caused by a 100 Hz current harmonic using a high load current and a Z-source capacitor of C_Z = 100 μF.

Section 2 presents the basic structures of impedance networks and calculations of the minimum ZSI output current $I_{OUTrmsmin}$ that ensure their continuous current mode (CCM). In Section 3 the idea of the inverter with the impedance network charging the battery from the DC link (during non-shoot-through time) to keep the impedance network in CCM is presented. The simulations and results of the experimental verification are presented. Section 4 contains the discussion of what kind of previously presented types of ZSI output voltage distortions can be canceled by the controlled charging of the battery. Section 5 presents the final conclusions.

2. Basic Impedance Networks: Z-Source and qZ-Source

The Z-source and qZ-source impedance networks shown in Figures 1 and 2, respectively, can operate in different states. Two basic states were taken into account during analysis and these include the shoot-through and the non-shoot-through states. The non-shoot-through state is depicted in Figures 1a and 2a, while the shoot-through state [23,27,28] is shown in Figures 1b and 2b.

The Z-source has a symmetrical structure where the values of the inductors are equal i.e., L_{Z1} = L_{Z2}. Similarly, the values of capacitors are the same, i.e., C_{Z1} = C_{Z2}, and the currents in both inductors are the same, i.e., i_{LZ1} = i_{LZ2}. In the qZ-source, the currents in both coils are the same and are identical to the Z-source coils currents (neglecting the influence of the different parasitic resistances) if coils have equal inductances.

The amplitude of the VSI output voltage V_{OUTmax} for the ZSI and qZSI is defined in Equation (1) as

$$V_{OUT\max} = \eta k'_V M V_{DC} = \eta \frac{M}{1-2d_Z} V_{DC} \quad (1)$$

where η is the efficiency, V_{DC} is the input voltage, M is the VSI modulation coefficient, and k_V' is the DC voltage boost factor of the impedance network without power losses [23,27,28].

It is assumed that the capacitance C_Z in the Z-source and qZ-source networks are sufficiently high. The average voltage on the capacitors of the Z-source and the C_{Z2} capacitor of the qZ-source are identical to the average voltage V_{LZav} on the inductors [23,27,28] given in Equation (2) as follows:

$$V_{LZ1av} = V_{LZ2av} = V_{LZ} = \frac{1-d_Z}{1-2d_Z}V_{DC} \tag{2}$$

The input power P_{IN} and output power P_{OUT} of the VSI connected to the impedance networks for a Z-source or qZ-source can be calculated using Equations (3)–(5):

$$P_{IN} = V_{DC}I_{DCav} = V_{DC}I_{LZav} \tag{3}$$

$$P_{OUT} = V_{OUTrms}I_{OUTrms} = \eta P_{IN} \tag{4}$$

$$P_{OUT} = \frac{1}{\sqrt{2}}\eta\frac{M}{1-2d_Z}V_{DC}I_{OUTrms} = \eta V_{DC}I_{LZav} \tag{5}$$

where I_{LZav} is a single inductor current averaged over the fundamental period T_m.

For the simplest case of the resistive ZSI load, R_{LOAD} the output power can be defined Equation (6) as

$$P_{OUT} = \left(\frac{1}{\sqrt{2}}\eta\frac{M}{1-2d_Z}V_{DC}\right)^2\frac{1}{R_{LOAD}} = \eta V_{DC}I_{LZav} \tag{6}$$

And the average inductor current I_{LZav} for the root mean square (rms) value of the inverter output current I_{OUTrms} is given Equation (7) as

$$I_{LZav} = \frac{1}{\sqrt{2}}\frac{M}{1-2d_Z}I_{OUTrms} \tag{7}$$

The i_{LZ} inductor current illustrated in Figure 4a comprises three components. These components are the average current I_{LZav}, the current i_{LZ2fm} which is averaged in the T_s switching period, and the triangle component $i_{LZ\Delta}$ of the inductor current. The current i_{LZ2fm} has the double fundamental frequency caused by the envelope of the input current of the VSI bridge in the non-shoot-through time while the triangle component inductor current $i_{LZ\Delta}$ is caused by storing energy in the coil during the shoot-through time and recovering energy in the rest of the switching period (in CCM). A plot of the VSI input current is displayed in Figure 4b.

The inductor current i_{LZ} is defined in Equation (8) as

$$i_{LZ}(t) = I_{LZav} + i_{LZ2fm}(t) + i_{LZ\Delta}(t) \tag{8}$$

Figure 4 shows plots of a Z-source or qZ-source impedance network coil current and an inverter input current including shoot-through current pulses for cases of maximum and close to zero crossing of the inverter output voltage (in CCM).

This most important harmonic component $2f_m$ of the VSI bridge input current flows through the L_ZC_Z circuit of the impedance network as shown in Equation (9). It is assumed that all power losses are within the impedance network including the power losses on the VSI switches during the shoot-through time.

$$i_{LZh2fm}(abs(i_{LOAD}(t))) = \frac{4}{3\pi}\sqrt{2}I_{OUTrms}\cos(4\pi f_m t)\left|\frac{1}{1-(4\pi f_m)^2L_ZC_Z}\right| \tag{9}$$

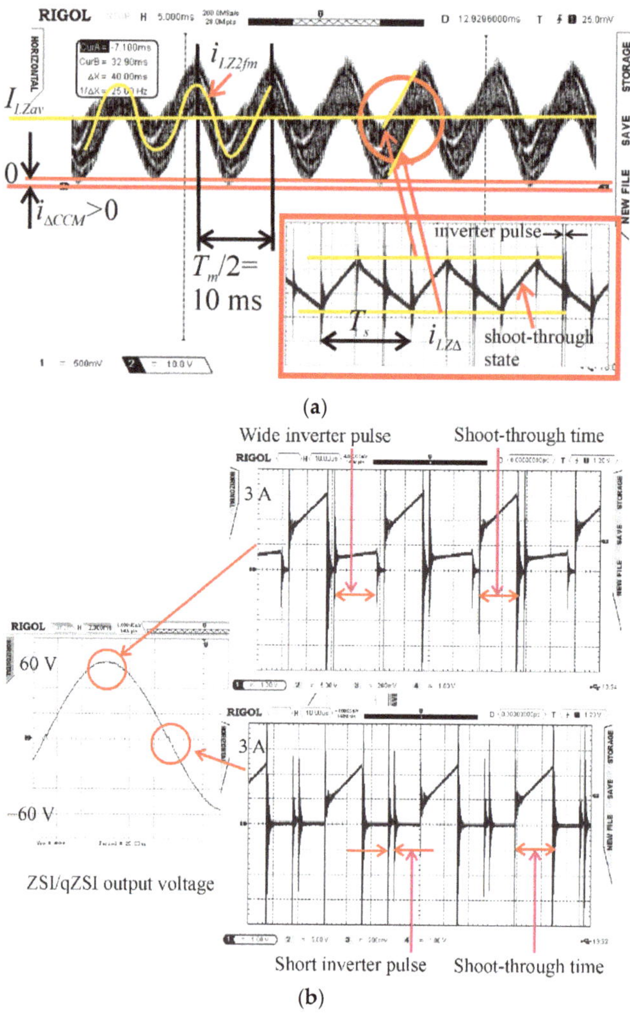

Figure 4. A Z-source or qZ-source impedance network (**a**) coil current and (**b**) the VSI input current including shoot-through current pulses (that do not supply inverter) in the case of wide (for the maximum of the output inverter voltage) and short (close to zero crossing of the output inverter voltage) inverter PWM pulses in the CCM.

The triangle component $i_{LZ\Delta}$ of the inductor current i_{LZ} in the CCM is calculated approximately with the assumption that a sufficiently low capacitor voltage ripple ΔV_{CZ} is approximately equal to 0 and V_{CZmax} is nearly equal to V_{CZav} for the shoot-through time. The triangle component $i_{LZ\Delta}$ can thus be expressed in Equation (10) as

$$i_{LZ\Delta}(t) \approx \frac{V_{CZav}}{L_Z}t,\ i_{LZ\Delta\max} = \frac{V_{CZav}}{L_Z}T_{st} = \frac{1}{L_Z}\frac{1-d_Z}{1-2d_Z}V_{DC}d_Z T_s,\ i_{LZ\Delta\max} = \sqrt{2}\frac{1}{L_Z}\frac{1-d_Z}{\eta M}V_{OUTrms}d_Z T_s \tag{10}$$

Consequently, the inductor current can be defined Equation (11) as

$$i_{LZ}(t) = [\frac{1}{2}\frac{M}{1-2d_Z} + \frac{4}{3\pi}\sqrt{2}\cos(4\pi f_m t)\left|\frac{1}{1-(4\pi f_m)^2 L_Z C_Z}\right|]I_{OUTrms} + i_{LZ\Delta}(t) \tag{11}$$

The lowest value of the inductor current is calculated Equation (12) as

$$i_{LZ\min}(t) = [\frac{1}{2}\frac{M}{1-2d_Z} - \frac{4\sqrt{2}}{3\pi}\left|\frac{1}{1-(4\pi f_m)^2 L_Z C_Z}\right|]I_{OUTrms} - \frac{1}{2}i_{LZ\Delta\max} \tag{12}$$

As shown in Figure 4a, the requirement for CCM is that i_{LZmin} must be greater than 0. This phenomenon is expressed in Equation (13) as

$$[\frac{1}{2}\frac{M}{1-2d_Z} - \frac{4\sqrt{2}}{3\pi}\left|\frac{1}{1-(4\pi f_m)^2 L_Z C_Z}\right|]I_{OUTrms} - \frac{1}{\sqrt{2}}\frac{1}{L_Z}\frac{1-d_Z}{\eta M}V_{OUTrms}d_Z T_s > 0 \tag{13}$$

From Figure 5a, the absolute value of load impedance expressed in Equation (14) should be lower in value (but always positive) than the value calculated in Equation (14) for CCM for the assigned parameters: d_Z, L_Z, and C_Z, $M = 1 - d_Z$.

$$|Z_{LOAD}| < \frac{\eta M L_Z}{(1-d_Z)d_Z T_s}(\frac{M}{\sqrt{2}}\frac{1}{1-2d_Z} - \frac{8}{3\pi}\left|\frac{1}{1-(4\pi f_m)^2 L_Z C_Z}\right|) \tag{14}$$

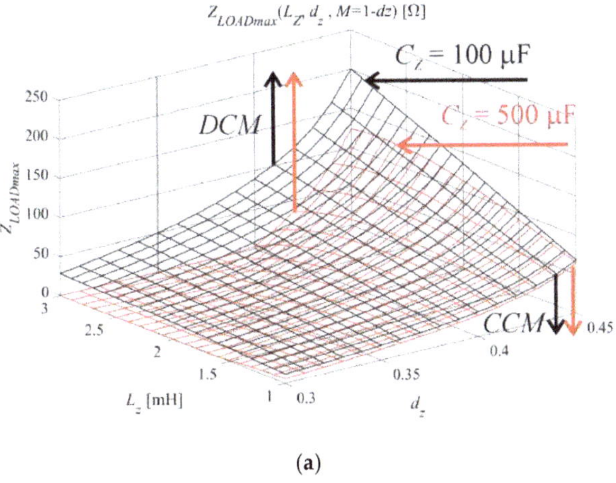

(a)

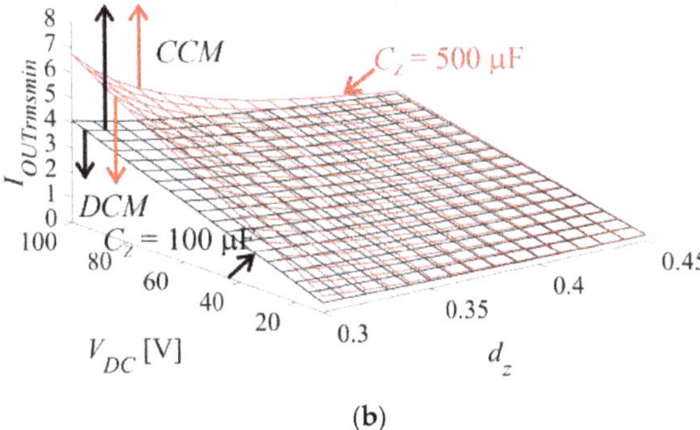

(b)

Figure 5. (**a**) Maximum load impedance, and (**b**) minimum output current, that keeps the impedance network in the continuous current mode.

As shown in Figure 5b, the minimum output current for CCM is given Equation (15) as

$$I_{OUTrms} > \frac{\frac{1}{ML_Z}\frac{1-d_Z}{1-2d_Z}V_{DC}d_Z T_s}{\frac{1}{1-2d_Z} - \frac{8\sqrt{2}}{M3\pi}\left|\frac{1}{1-(4\pi f_m)^2 L_Z C_Z}\right|} \quad (15)$$

The impedance network (Figure 5b) operates in the CCM for the ZSI load current I_{OUTrms} higher than the value calculated from Equation (15) for assigned L_Z = 1 mH and three parameters: V_{DC}, d_Z, and C_Z. The modulation index M has the assigned maximum possible value $M = 1 - d_Z$.

In Figure 6, the continuous current mode is illustrated where the output voltage of the ZSI is undistorted.

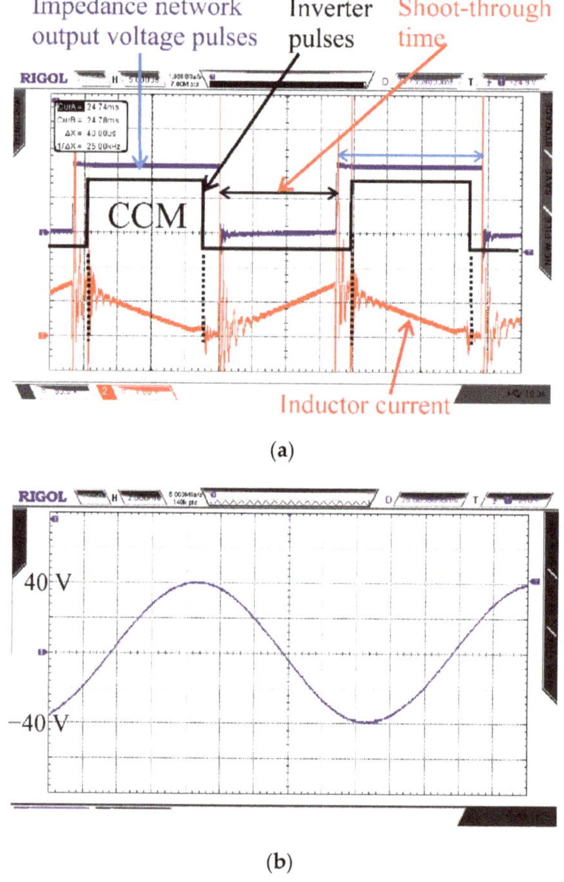

Figure 6. CCM waveforms of (**a**) the I_{LZ} coil current, ZSI output voltage, and inverter PWM pulses, and (**b**) the undistorted inverter output voltage.

Figure 7 presents the DCM where two cases can be distinguished. From this figure, the distortions of the output voltage are small when the output voltage is below the maximum. When the output voltage is closer to the maximum, the distortions are higher, and the output voltage maximum is lower than expected. For the large VSI output capacitor the VSI output and PWM envelope voltages are shifted when the large VSI output capacitor e.g., C_F = 50 µF is used. As shown in Figure 7, the short PWM pulses are undistorted in DCM while the wide pulses are distorted, and the output voltage is lower. The simulation of a DCM operation using the Z-source is presented in Figure 8 for the third PWM modulation schema [30]. The variables used to obtain the measured plots in Figure 8 are given as: C_F = 1 µF, d_Z = 0.3, M = 0.65, R_{LOAD} = 1000 Ω, 3rd modulation schema.

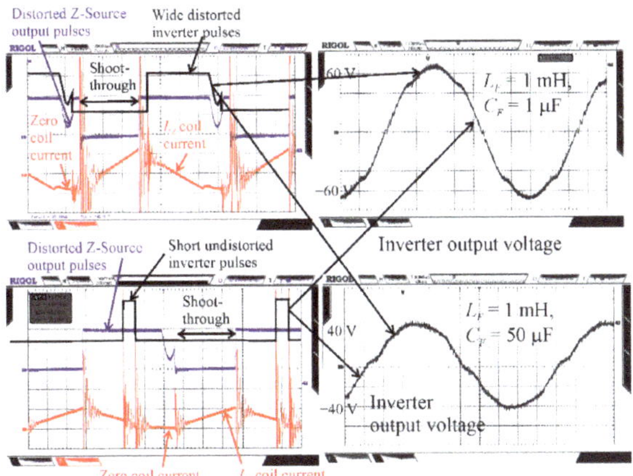

Figure 7. Measured DCM waveforms of the L_Z coil current, ZSI output voltage, and the inverter's PWM wide and short pulses for C_F = 1 µF and 50 µF inverter capacitors.

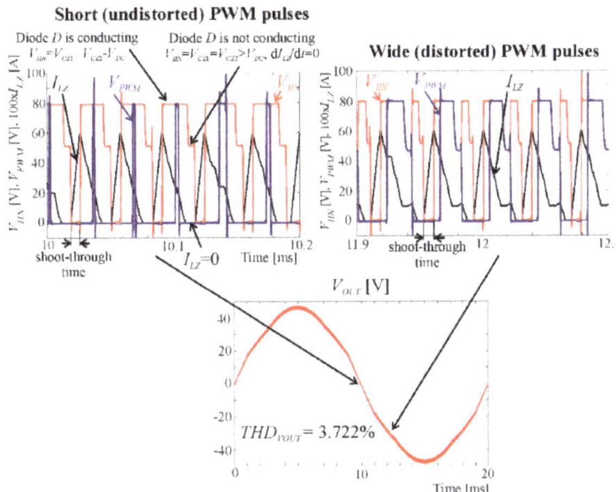

Figure 8. Simulated DCM waveforms for inverter C_F = 1 µF, d_Z = 0.3, M = 0.65, R_{LOAD} = 1000 Ω, 3rd modulation schema.

3. Controlled Energy Flow—Charging the Battery

Similar results of measurement shown in Figure 7 and simulations in Figure 8 demonstrate that further simulations of the controlled energy flow i.e., charging the battery is useful. The basic solution is an efficient multi-input-single-output (MISO) [31] feedback that can decrease total harmonic distortions (*THD*) [23,27]. In addition, MISO feedback can decrease two other types of output voltage distortions [27]. However, for systems supplied by varying the DC supply voltage, for example, photovoltaic cells, the controlled energy flow to the batteries, which keeps the CCM, can be used. It is recommended that the battery is charged with a current that is a function of the difference between the calculated value of $I_{OUTrmsmin}$ and averaged (10 Hz low pass filter) VSI output current I_{OUTrms} as shown in

Figure 9 (if this difference is negative the charging battery current is equal to zero). The actual difference of these currents $I_{OUTrmsmin} - I_{OUTrms}$ is recalculated (if positive) to match the required increase of the average I_{LZav} current expressed in Equation (7). The battery can be charged only during the non-shoot-through state. Energy from the battery is discharged when V_{DC} decreases below the assumed value of V_{DCmin}, the Z-source is switched off and the shoot-through pulses are blocked.

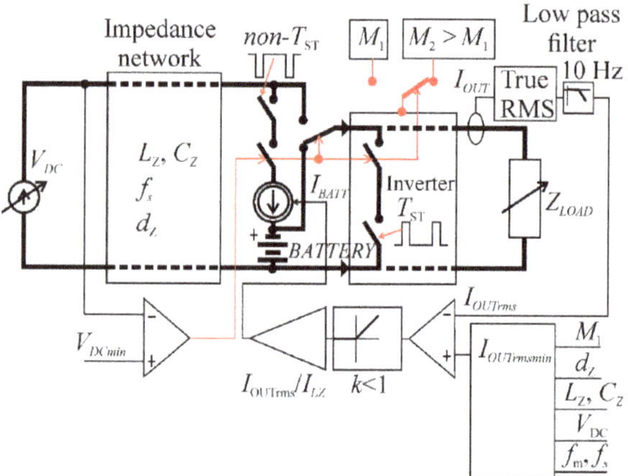

Figure 9. Proposed idea of the inverter with the impedance network charging the battery from the DC link (during non-shoot-through time), and automatic switching to supplying directly from the battery (the positions of switches are presented in the position of discharging the battery when $V_{DCmin} - V_{DC} > 0$).

The idea of this system is presented in Figure 9 (for switches placed in the position of discharging the battery). When the battery returns energy, the following happens: the shoot-through pulses are stopped, and the 48 V battery is connected directly to the VSI. This battery voltage should be higher than the amplitude of the output sinusoidal voltage and the modulation index M of VSI is increased i.e., M_2 is greater than M_1 (Figure 9).

Figure 10a presents the simulated waveforms of the V_{DC} changed 24/12/24 V (the border value is set to 15 V) with the described automatic action from Figure 9 but without controlled charging the battery when Z-Source operates in the DCM. The following parameters were used in this scenario: $d_Z = 0.3$, $M_1 = 0.65$, $M_2 = 0.75$ and $R_{LOAD} = 1000\ \Omega$. Figure 10b presents that same operation but with controlled charging of the battery for keeping Z-Source in the CCM. The current charging of the battery is calculated as $I_{BATT} = f(I_{OUTrmsmin} - I_{OUTrms})$ using Equation (15), where f is a function of Equation (7). The battery charging current I_{BATT} calculated from Equations (7) and (15) should be reduced because too high a value of the battery charging current leads to distortions of the VSI output voltage time after the output voltage is zero-crossing (see Figure 11b). These distortions are caused by the high voltage drops on the VSI switches during the shoot-through time. The presented (Figure 10b) reduction of the output voltage THD from 4.6% to 3% without any feedback loop is quite promising.

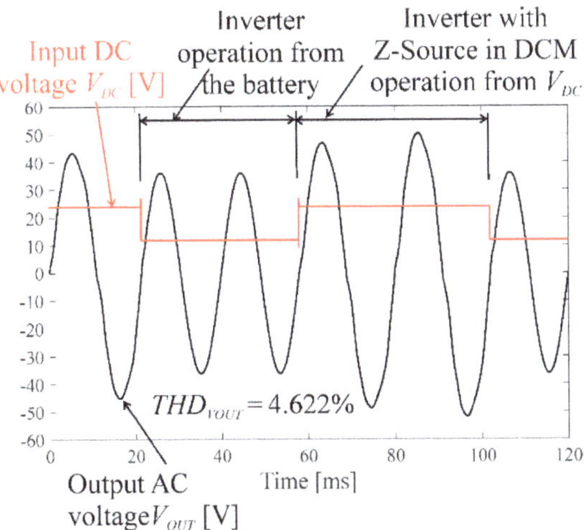

(a)

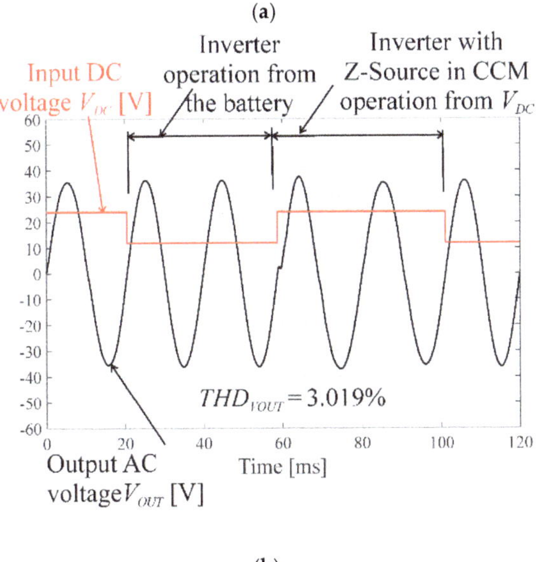

(b)

Figure 10. The waveforms of the DC input and AC output voltages of the ZSI switched from a mode of supplying the VSI from Z-source to the mode of supplying VSI from the battery in case of the low input DC voltage, (**a**) without controlled charging battery for Z-source in the DCM for the low load, and (**b**) with controlled charging battery for Z-source in CCM.

The presented simulations were verified in an experimental model using a 12 V battery (without discharging the battery) charged from the DC during $d_B T_s$ pulses where ($d_B = 1 - d_Z$) (Figure 12). The feedback loop was the IPBC2 type presented in [27]. For the DCM mode of the Z-source, the output voltage distortions can be reduced by additional loading the impedance network by means of charging the battery from the DC link in the non-shoot through times.

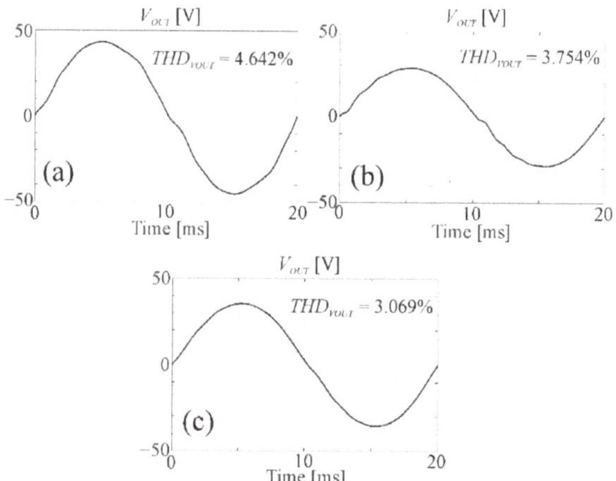

Figure 11. Inverter output voltage (**a**) without charging battery, (**b**) the battery charging current directly equal to $f(I_{OUTrmsmin} - I_{OUTrms})$, where f is a Equation (7), and (**c**) the battery charging with the reduced value of current.

(**a**)

Figure 12. *Cont.*

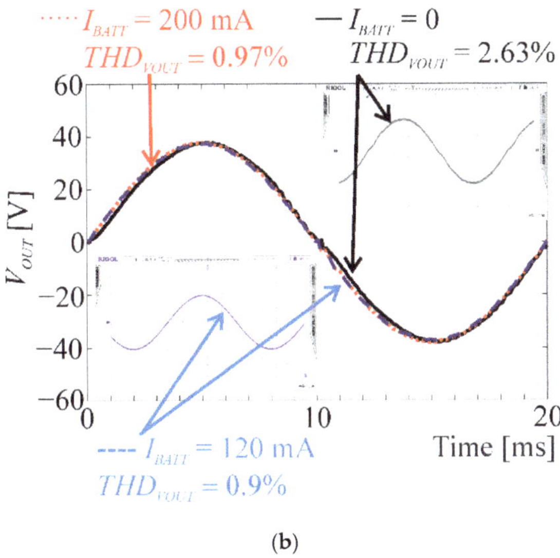

(**b**)

Figure 12. (**a**) The inverter experimental set up and (**b**) inverter output voltage distortions comparison for an IPBC controller where R_{LOAD} = 2000 Ω, RMS battery charging currents: I_{BATT} = 0 (DCM of the Z-source), I_{BATT} = 120 mA and I_{BATT} = 200 mA (CCM of the Z-source), d_Z = 0.3, and $d_B = 1 - d_Z$—battery charging pulses coefficient.

The current source from Figure 9 was simply substituted with resistors. Charging the battery allowed for a substantial reduction of output voltage THD from 2.63% to 0.9%. for I_{BATT} = 120 mA, but THD increased to 0.97% for I_{BATT} = 200 mA. Further research will be on the use of battery charging current not only to reduce the distortions of the output voltage but also looking for a maximum power point (MPP) when the impedance network is supplied from the photovoltaic cell. The battery charging current can be controlled by the coefficient d_B for the input current of the impedance network would be closer to MPP.

4. Discussion

The presented results of the simulation and measurements of the experimental ZSI proved that charging the battery from the DC link between impedance network and VSI in the non-shoot-through time can seriously decrease the ZSI output voltage distortions keeping the impedance network in the CCM. The controlled energy flow solution is particularly predicted for the case of wide variations of the input DC voltage and variations of the load current. The output voltage distortions are decreased even when a strong feedback loop of the VSI is present. The controlled charging of the battery can help in the maximum power point tracking when the ZSI is supplied from the photovoltaic cell and this is the perspective of the further studies. In [23], three types of VSI output voltage distortions were distinguished. The controlled charging of the battery can cancel one of them but setting too high a value of this current increases the other reason for distortions. Charging the battery from the DC link of the ZSI during the non-shoot-through time was not presented yet, however, another approach to the controlled power flow for qZSI with charging the battery connected parallel to the C_{Z2} capacitor (Figure 2) was presented in [32].

5. Conclusions

In this paper, a technique has been proposed to reduce output voltage distortions in voltage source inverters connected to impedance networks. The proposed method has been validated using simulations and experimentally under different operating conditions. It was discovered that by connecting a rechargeable battery to a DC link placed between an impedance network and a VSI and employing proper control of the battery charging current during the non-shoot through time, the output voltage distortions in a system with or without feedback can be reduced when a continuous current mode of the impedance network is forced. However, too high a current charging the battery may increase other types of VSI output voltage distortions presented in Figure 11b caused by high voltage drops on the VSI switches during the shoot-through time. Furthermore, the battery charging current can be controlled to increase the impedance network input current to enable the system to reach the maximum power point when the DC source is a photovoltaic cell. The results presented in this paper thus demonstrate that the proposed method is suitable and can be applied in practice to real-time supply systems.

Author Contributions: Conceptualization, Z.R.; methodology, Z.R. and K.B.; software, Z.R.; validation, Z.R., K.B. and Ł.D.; formal analysis, Z.R. and K.B.; investigation, Z.R. and K.B.; resources, Z.R. and K.B.; data curation, Z.R. and K.B.; writing—original draft preparation, Z.R.; writing—review and editing, Z.R. and K.B.; visualization, Z.R.; supervision, Z.R.; project administration, Z.R. and K.B.; funding acquisition, Z.R. and K.B. All authors have read and agreed to the published version of the manuscript.

Funding: This work was partly supported by the Polish Ministry of Science and Higher Education funding for statutory activities.

Conflicts of Interest: The authors declare no conflict of interest. The funders had no role in the design of the study; in the collection, analyses, or interpretation of data; in the writing of the manuscript, or in the decision to publish the results.

References

1. Peng, F.Z. Z-Source Inverter. *IEEE Trans. Ind. Appl.* **2003**, *39*, 504–510. [CrossRef]
2. Anderson, J.; Peng, F. Four quasi-Z-Source inverters. *Proc. IEEE Power Electron. Spec. Conf.* **2008**, 2743–2749. [CrossRef]
3. Li, Y.; Peng, F.Z. AC Small Signal Modeling, Analysis and Control of Quasi-Z-Source Converter. In Proceedings of the 2012 IEEE 7th International Power Electronics and Motion Control Conference—ECCE Asia, Harbin, China, 2–5 June 2012; pp. 1848–1854. [CrossRef]
4. Liu, W.; Yang, Y.; Kerekes, T.; Liivik, E.; Blaabjerg, F. Impedance Network Impact on the Controller Design of the QZSI for PV Applications. In Proceedings of the 2020 IEEE 21st Workshop on Control and Modeling for Power Electronics (COMPEL), Aalborg, Denmark, 9–12 November 2020. [CrossRef]
5. Subhani, N.; Kannan, R.; Mahmud, A.; Blaabjerg, F. Z-source inverter topologies with switched Z-impedance networks: A review. *IET Power Electron.* **2021**, *14*, 727–750. [CrossRef]
6. Zhu, M.; Yu, K.; Luo, F.L. Switched Inductor Z-Source Inverter. *IEEE Trans. Power Electron.* **2010**, *25*, 2150–2158. [CrossRef]
7. Itozakura, H.; Koizumi, H. Embedded Z-Source Inverter with Switched Inductor. In Proceedings of the IECON 2011—37th Annual Conference of the IEEE Industrial Electronics Society, Melbourne, VIC, Australia, 7–10 November 2011; pp. 1342–1347. [CrossRef]
8. Adamowicz, M.; Guzinski, J.; Strzelecki, R.; Peng, F.Z.; Abu-Rub, H. High Step-Up Continuous Input Current LCCT-Z-Source Inverters for Fuel Cells. In Proceedings of the Energy Conversion Congress and Exposition (ECCE), Phoenix, AZ, USA, 17–22 September 2011; pp. 2276–2282. [CrossRef]
9. Adamowicz, M. LCCT-Z-Source Inverters. In Proceedings of the 10th International Conference on Environment and Electrical Engineering, Rome, Italy, 8–11 May 2011; pp. 1–6. [CrossRef]
10. Vinnikov, D.; Roasto, I.; Strzelecki, R.; Adamowicz, M. Step-Up DC/DC Converters With Cascaded Quasi-Z-Source Network. *IEEE Trans. Ind. Electron.* **2012**, *59*, 3727–3736. [CrossRef]
11. Siwakoti, Y.P.; Blaabjerg, F.; Galigekere, V.P.; Ayachit, A.; Kazimierczuk, M.K. A-Source Impedance Network. *IEEE Trans. Power Electron.* **2016**, *31*, 8081–8087. [CrossRef]
12. Hakemi, A.; Sanatkar-Chayjani, M.; Monfared, M. Δ-Source Impedance Network. *IEEE Trans. Ind. Electron.* **2017**, *64*, 1–10. [CrossRef]
13. Rezazadeh, H.; Monfared, M.; Nikbahar, A.; Sharifi, S. A family of high voltage gain quasi-Δ-source impedance networks. *IET Power Electron.* **2021**, *14*, 807–820. [CrossRef]
14. Siwakoti, Y.P.; Peng, F.Z.; Blaabjerg, F.; Loh, P.C.; Town, G.E. Impedance-Source Networks for Electric Power Conversion Part I: A Topological Review. *IEEE Trans. Power Electron.* **2015**, *30*, 699–716. [CrossRef]
15. Reddivari, R.; Jena, D. A Correlative Investigation of Impedance Source Networks: A Comprehensive Review. *IETE Tech. Rev.* **2021**, *38*, 1–34. [CrossRef]
16. Ghasimi, S.; Eshkevari, A.L.; Mosallanejad, A. A high-gain IΓ-source hybrid single-phase multilevel inverter for photovoltaic application. *IET Power Electron.* **2021**, *14*, 106–119. [CrossRef]

17. Kumar, A.; Bao, D.; Beig, A.R. Comparative Analysis of Extended SC-qSBI with EB-QZSI and EB/ASN-QZSI. *IEEE Access* **2021**, *9*, 61539–61547. [CrossRef]
18. Zhao, P.; Wang, J.; Hao, H.; Wang, U. Y-Source Two-Stage Matrix Converter and Its Modulation Strategy. *IEEE Access* **2020**, *8*, 214282–214292. [CrossRef]
19. Yuan, J.; Mostaan, A.; Yang, Y.; Siwakoti, Y.P.; Blaabjerg, F. A Modified Y-Source DC–DC Converter With High Voltage-Gains and Low Switch Stresses. *IEEE Trans. Power Electron.* **2020**, *35*, 7716–7720. [CrossRef]
20. Gayen, P.K. An enhanced high-boost active-switched quasi Z-Source inverter having shorter range of shoot-through duty ratio for solar energy conversion applications. *AEU—Int. J. Electron. Commun.* **2021**, *137*, 153822. [CrossRef]
21. Siwakoti, Y.P.; Peng, F.Z.; Blaabjerg, F.; Loh, P.C.; Town, G.E.; Yang, S. Impedance-Source Networks for Electric Power Conversion Part II: Review of Control and Modulation Techniques. *IEEE Trans. Power Electron.* **2015**, *30*, 1887–1906. [CrossRef]
22. Abdelhakim, A.; Blaabjerg, F.; Mattavelli, P. Modulation Schemes of the Three-Phase Impedance Source Inverters—Part I: Classification and Review. *IEEE Trans. Ind. Electron.* **2018**, *65*, 6309–6320. [CrossRef]
23. Rymarski, Z.; Bernacki, K. Drawbacks of impedance networks. *Int. J. Circuit Theory Appl.* **2018**, *46*, 612–628. [CrossRef]
24. Bernacki, K.; Rymarski, Z. Electromagnetic Compatibility of Impedance Source Inverters. *Elektron. Elektrotechnika* **2017**, *23*, 55–63. [CrossRef]
25. Hufman, B. Efficiency and Power Characteristics of Switching Regulator Circuits. *Linear Technol. Appl. Note* **1991**, *46*, 1–28.
26. Ge, B.; Abu-Rub, H.; Peng, F.Z.; Lei, Q.; Almeida, A.T.; Ferreira, F.J.T.E.; Sun, D.; Liu, Y. An Energy-Stored Quasi-Z-Source Inverter for Application to Photovoltaic Power System. *IEEE Trans. Ind. Electron.* **2013**, *60*, 4468–4481. [CrossRef]
27. Rymarski, Z.; Bernacki, K.; Dyga, Ł. Decreasing the single phase inverter output voltage distortions caused by impedance networks. *IEEE Trans. Ind. Appl.* **2019**, *55*, 7586–7594. [CrossRef]
28. Rymarski, Z.; Bernacki, K. Influence of Z-Source output impedance on dynamic properties of single-phase voltage source inverters for uninterrupted power supply. *IET Power Electron.* **2014**, *7*, 1978–1988. [CrossRef]
29. Bernacki, K.; Rymarski, Z.; Dyga, Ł. Selecting the coil core powder material for the output filter of a voltage source inverter. *Electron. Lett.* **2017**, *53*, 1068–1069. [CrossRef]
30. Bernacki, K.; Rymarski, Z. Electromagnetic compatibility of voltage source inverters for uninterruptible power supply system depending on the pulse-width modulation scheme. *IET Power Electron.* **2015**, *8*, 1026–1034. [CrossRef]
31. Astrom, K.J.; Wittenmark, B. *Computer-Controlled Systems: Theory and Design*, 3rd ed.; Dover Publications Inc.: Mineola, NY, USA, 2011; ISBN 9780486486130.
32. Sun, D.; Ge, B.; Peng, F.Z.; Abu Rub, H.; de Almeida, A.T. Power flow control for quasi-Z source inverter with battery based PV power generation system. In Proceedings of the 2011 IEEE Energy Conversion Congress and Exposition, Phoenix, AZ, USA, 17–22 September 2011; pp. 1051–1056. [CrossRef]

Review

A Comprehensive Review of Lithium-Ion Cell Temperature Estimation Techniques Applicable to Health-Conscious Fast Charging and Smart Battery Management Systems

Akash Samanta [1,*] and Sheldon S. Williamson [2]

[1] Department of Applied Physics, Faculty of Electrical Engineering, University of Calcutta, Kolkata 700009, India
[2] Department of Electrical, Computer and Software Engineering, Faculty of Engineering and Applied Science, Ontario Tech University, Oshawa, ON L1G 0C5, Canada; Sheldon.Williamson@ontariotechu.ca
* Correspondence: akashsamanta440@gmail.com; Tel.: +91-9143877405

Abstract: Highly nonlinear characteristics of lithium-ion batteries (LIBs) are significantly influenced by the external and internal temperature of the LIB cell. Moreover, a cell temperature beyond the manufacturer's specified safe operating limit could lead to thermal runaway and even fire hazards and safety concerns to operating personnel. Therefore, accurate information of cell internal and surface temperature of LIB is highly crucial for effective thermal management and proper operation of a battery management system (BMS). Accurate temperature information is also essential to BMS for the accurate estimation of various important states of LIB, such as state of charge, state of health and so on. High-capacity LIB packs, used in electric vehicles and grid-tied stationary energy storage system essentially consist of thousands of individual LIB cells. Therefore, installing a physical sensor at each cell, especially at the cell core, is not practically feasible from the solution cost, space and weight point of view. A solution is to develop a suitable estimation strategy which led scholars to propose different temperature estimation schemes aiming to establish a balance among accuracy, adaptability, modelling complexity and computational cost. This article presented an exhaustive review of these estimation strategies covering recent developments, current issues, major challenges, and future research recommendations. The prime intention is to provide a detailed guideline to researchers and industries towards developing a highly accurate, intelligent, adaptive, easy-to-implement and computationally efficient online temperature estimation strategy applicable to health-conscious fast charging and smart onboard BMS.

Keywords: electric vehicles; machine learning; Kalman filter; thermal modelling; online prediction; electromagnetic impedance spectroscopy; computational cost

Citation: Samanta, A.; Williamson, S.S. A Comprehensive Review of Lithium-Ion Cell Temperature Estimation Techniques Applicable to Health-Conscious Fast Charging and Smart Battery Management Systems. *Energies* **2021**, *14*, 5960. https://doi.org/10.3390/en14185960

Academic Editors: Andrei Blinov, Sheldon Williamson, Seung-Wan Song and Mario Marchesoni

Received: 5 July 2021
Accepted: 13 September 2021
Published: 20 September 2021

1. Introduction

Lithium-ion batteries (LIBs) are widely used in electric vehicles (EVs), grid-tied stationary energy storage systems, and several other consumer electronics primarily due to their high voltage rating (>4 V/cell) and high energy density (~265 (W h) L^{-1}) and longer operational life. The use of LIBs in automotive and aerospace applications has led to larger cell sizes and large battery packs for a higher driving range and the requirement for more aggressive charging and discharging. However, thermal instability and temperature-dependent nonlinear behavior is some of the common concerns behind the safe and reliable operation of LIB systems. It is noticed that the operation of batteries outside the safe operating temperature directly affects the performance of LIBs, such as cycle life, efficiency, reliability and safety. Researchers investigating the thermal performance of LIB showed that the best operating temperature range is from 25 °C to 40 °C [1,2]. Richardson et al. [3] demonstrated that the difference between the core and surface temperature could reach more than 10 °C during real-life applications, especially during the high discharging condition and fluctuating load current demand. The excessive temperature difference and the accumulation of a large amount of heat inside the cell could lead to thermal runaway or even explosions and fire [4]. That necessitates the employment of a battery management system (BMS) for effective monitoring of battery parameters (current, voltage, temperature), estimation of battery states (state of charge (SOC), state of health (SOH), remaining useful life (RUL), state of temperature (SOT) [5]). Research studies demonstrated that SOC [6], SOH [7], and remaining storage capacity [8] are a function of temperature; thus, the estimation of the battery states also depends on the accurate estimation of cell temperature. The Columbic efficiency of a cell is greatly affected by the cell temperature during the charging and discharging period. Few other popular functionalities of BMS include cell balancing [9] and fault detection/diagnosis [10] to ensure optimum capacity utilization, operational safety, reliability, and longer battery life often requires temperature information of an individual cell and battery pack as well. Therefore, accurate information of core and surface temperature is highly crucial for effective thermal management and safety of a LIB pack. Moreover, in cold climate areas, the battery capacity is drastically reduced due to low-temperature operation that requires preheating the battery to a suitable range for optimum performance [11,12]. It is also evidenced that for every 0.1 °C beyond the safe operating region the battery capacity degrades by about 5% [13]. It is evidenced that maximum heat is generated during the discharging period especially with fast discharging [14]. Therefore, accurate temperature estimation is essential for effective thermal management and safety during fast charging and discharging and preheating of the cell to minimize capacity fade.

In summary, it could be stated that the accurate information of cell temperature is undoubtedly serving as the essential basis for the thermal management and safety of LIB. While the surface temperature of each cell can be measured by installing a temperature sensor on each cell, the core or internal temperature measurement directly using physical sensors is challenging. Moreover, installing a temperature sensor on each cell surface is not practically feasible from a system cost, space and weight point of view as any high-capacity battery pack used in EVs and grid-tied systems essentially consists of thousands of individual cells. Researchers have also incorporated multi-dimensional sensing and self-healing functions into a single battery cell to develop a smart battery [15–18]. Smart cells are typically capable of parameter measurements and estimation of cell states including the state of temperature. Despite the modularized application of BMS in smart batteries, accurate temperature estimation is still required, as otherwise installing sensors in each cell results in high implementation cost and complexity. Therefore, researchers are struggling hard to develop a high-fidelity, accurate, easy-to-implement, and computationally inexpensive online temperature estimation strategy suitable for low-cost onboard BMS. Several temperature estimation techniques have been proposed by researchers so far. Each different type of method has its advantages and limitations with respect to the above-mentioned features of an optimum BMS. Therefore, a summary of all the prominent techniques would be very helpful to researchers and developers serving as a baseline for further research and as a guideline for selecting appropriate techniques suitable for a specific requirement. However, such a summary with detailed discussion on current progress and explanation of the existing issues, challenges and future research scopes has not yet been presented in the literature. Therefore, this article covered the research gap by conducting a comprehensive review of the state-of-the-art temperature estimation strategies reported in the literature so far.

The paper is organized as follows: In Section 2, generic temperature estimation strategy of LIB is presented. The classification of temperature estimation strategies is presented in Section 3. Section 4 is dedicated to presenting the existing estimation techniques, their evolutions, limitations and challenges. It should be noted that temperature estimation strategies for LIBs reported in the literature between 2010 to 2021 are primarily considered. However, few prominent research articles published between 1990 to 2010 are also considered for understanding the fundamentals and evolution of temperature estimation schemes. Commonly used search platforms, such as "Google Scholar", "Science Direct", and IEEE Xplore, were used to find research articles published within this tenure. The search criteria were "Temperature Estimation of Lithium-ion Batteries". Section 5 discusses the current issues, challenges and future research recommendations. Finally, Section 6 is dedicated to a summary of the major findings and concluding remarks.

2. Generic Temperature Estimation Strategy

Irrespective of battery chemistry, heat is accumulated inside the battery during the charging/discharging even during idle conditions, majorly due to several largely exothermic chemical and electrochemical reactions as well as transport processes. If the heat transfer from the battery to the surroundings is not sufficient, then the heat gets accumulated inside the battery resulting in an increase in core and surface temperature, thereby risking thermal runaway. This phenomenon is even more prominent in the case of hard-cased insulated batteries (as used in EVs), under fast charging/discharging and the operation in hot environments. Heat dissipation is worse in cylindrical LIBs that are extensively used in high-capacity LIB packs. Therefore, a typical temperature estimation scheme consists of two models, namely, a heat generation model and a heat transfer model [19]. Often, a battery electrical model is also used to estimate the total heat generation using Bernardi's [20] heat generation model whereas few other models use a mathematical form of battery electrochemistry to calculate the heat generation. Adaptive estimation strategies also consider the influence of different battery states, such as SOC and SOH, as the battery temperature is a function of these battery states. Then, the heat transfer model takes the estimated total heat quantity as well as few other external measurements such as ambient temperature to predict the temperature of that cell. Closed-loop estimation schemes use the measured or the estimation temperature as feedback to improve the prediction accuracy. A schematic layout of a generic temperature estimation strategy for LIB is shown in Figure 1.

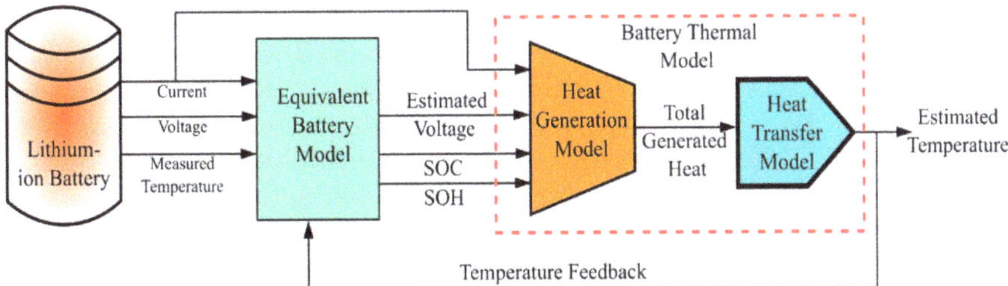

Figure 1. Schematic layout of a generic temperature estimation strategy for a LIB cell.

3. Classification of Temperature Estimation Strategies

As shown in Figure 1, typically, a temperature estimation scheme consists of a heat generation model and a heat transfer model. The heat generation models reported in the literature can be broadly classified from two different aspects; based on modelling strategy and based on the source of heat generation. Heat generation models based on modelling strategy can be classified into three groups, physics-based electrochemical models [21–24], equivalent circuit models (ECM) [25–27], black-box models [28–30]. In contrast, based on the source of heat generation, these models can be grouped as a concentrated model, distributed model [31] and heterogeneous model [25,32]. The concentrated heat generation model considers that all heat is generated at the core only, usually considered to reduce the modelling complexity. The distributed heat generation model considers that uniform heat is generated throughout the entire cell geometry whereas the heterogeneous model can capture different heat generation from difference cell layers usually resulting in temperature and current density gradients inside the cell. The heterogeneous models are more detailed thus can produce highly accurate predictions; however, these are most complex and require extensive experiments for modelling. Distributed heat generation models are a balance between the concentrated and heterogeneous models. The heat transfer models can be classified into finite element analysis (FEA)-based models [27,33–36], heat capacitor-resistor models (lumped or distributed parameter) [28,37–40], and data-driven techniques. Heat capacitor–resistor-based models use the analogy between electrical and thermal systems. A heat capacitor–resistor can be further classified as mentioned in Figure 2. Lumped parameter models are simple and useful for online applications, however, only one or two average temperatures can be predicted with these models whilst the battery temperature distribution is not spatially uniform, especially in larger capacity cylindrical LIB cells. On the other hand, complex distributed models [41,42] can describe the detailed temperature distribution in a cell, however, they are not suitable for online application due to their computational complexity. Several other detailed models of LIB accounting for the thermal characteristics of different layers are studied in [43–48]. A two-state/node model provides information on core and surface temperature whereas a one-state/node model can provide only core temperature.

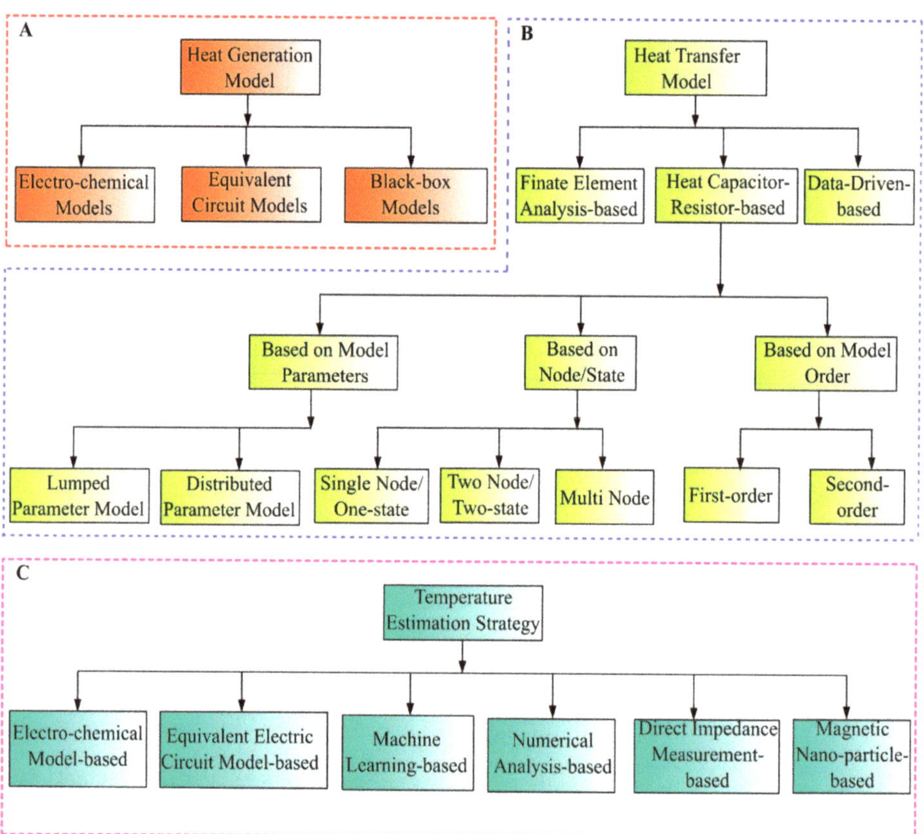

Figure 2. Family of (**A**) Heat generation model, (**B**) Heat transfer model, (**C**) Temperature estimation strategy.

The heat transfer model where the total heat generation is one of the input parameters is collectively called the battery thermal model where the total heat generation is estimated by the battery heat generation model. The thermal modelling of LIB is a separate area of study and is not under the scope of this study. It deals only with the temperature estimation strategies. However, as most of the temperature estimation strategies are extensively depending on thermal modelling, an overview of each modelling technique is also discussed with the respective temperature estimation strategy for better understanding. Researchers employed different types of heat generation models with different kinds of heat transfer models to come up with a temperature estimation scheme. Therefore, it is challenging to classify these estimation strategies. Broadly, the temperature estimation schemes can be grouped into electrochemical thermal modelling-based, equivalent electric circuit model (EECM)-based, machine learning (ML)-based, numerical-model based, direct impedance measurement-based, magnetic nanoparticles-based schemes. The families of the LIB heat generation model, heat transfer model and temperature estimation strategy are illustrated in Figure 2.

4. Comprehensive Review of Temperature Estimation Strategies

4.1. Electrochemical Thermal Modelling-Based Temperature Estimation

Researchers started thermal modelling in the early nineties, those are mostly coupled with an electrochemical model to simulate the temperature profile of a battery under different operating conditions, geometries or cooling rates. There are simple one-dimensional (radial direction) models [37,49–54] to complex three-dimensional thermal models [55–59]. Researchers have primarily used different analytical techniques to mathematically model the electrochemical behavior of the cell. One-dimensional models typically assume isothermal, constant current operation of the battery and lumped thermophysical properties and constant heat generation rates. Highly complex three-dimensional models require an in-depth understanding of the thermodynamic properties of battery materials and parts to consider the heat effects caused by ohmic resistance, chemical reactions, mixing processes, polarization and electrode kinetic resistance. Often, temperature estimation using such highly complex models is very accurate, however, such detailed models are essential for battery design purposes. Those are not compatible with temperature estimation using onboard BMS with low computational resources. These complex models are capable of accounting for the time-varying nonlinear battery performance. However, they typically require several system properties, operational parameters which require extensive experimental measurements. While, at the same time, quantitative estimation of some of the properties, such as transport properties, thermodynamic properties and heat effects are highly challenging.

Thomas and Newman [60] introduced an electrochemical modelling-based detailed heat generation model of LIB to estimate the total heat generation during the charging/discharging period. The fundamental equation of the total heat generation inside the LIB cell as proposed by Thomas and Newman reads

$$Q = I(V - U^{avg}) + IT\frac{\partial U^{avg}}{\partial T} - \sum_i \Delta H_i^{avg} r_i - \int \sum_j (\overline{H}_j - \overline{H}_j^{avg})\frac{\partial C_j}{\partial t} dv \quad (1)$$

In Equation (1), Q is the rate of heat generated or consumed inside the cell, V and U are the cell voltage and equilibrium potential, respectively, I is the charging or discharging current, T is the cell temperature. ΔH_i represents the changes in enthalpy of the chemical reaction i and r_i is the rate of reaction i. $\Delta \overline{H}_j$ represents the partial molar enthalpy of species j and c_j is the concentration of the species. t and v represent the time and volume of the cell, respectively. All the properties are mentioned based on the volume-averaged concentration, thus the superscript "*avg*" is used. The model can provide accurate information on heat generation only, temperature estimation is not presented in this study. Their heat generation model was extensively used by several other researchers. Modelling is very detailed thus highly complicated and not suitable for online application owing to the computational burden. One of the widely-used electrochemical models commonly known as the Doyle–Fuller–Newmann model [37,61] is extensively referred to and also used for thermal modelling. It consists of nonlinear partial differential-algebraic equations to describe the internal characteristics of LIB. It is also referred to as the pseudo-two-dimensional (P2D) model. The major limitation of the model is its high computational burden which limits its application in online state estimation in embedded BMSs. Here,

Al Hallaj et al. [53] showed that a simplified transient one-dimensional thermal model with lumped parameters is sufficient for cell design purposes, especially to simulate the thermal behavior of scaled-up LIBs. Detailed knowledge of the role of different cell components, such as electrodes, electrolytes and separators in heat generation is also not necessary. Few researchers used this type of complex electrochemical model to explore pulse power limitations to prevent thermal runaway and to design thermal management systems [62,63]. Those are mostly used for designing LIB cells as well as LIB packs. A lumped electrochemical-thermal-coupled model was used to predict the thermal performance of LIB alongside the performance of individual electrodes at various operating temperatures by Fang et al. [64]. The model was validated against the experimental data for constant current and pulsing conditions characteristic of hybrid electric vehicle (HEV) which are merely providing the laboratory experimental results instead of a real-work application scenario. The impact of charging current on internal temperature behavior was investigated in [65]. Gerver et al. [66] included more detailed information and cell characteristics to develop a multi-dimensional electrochemical thermal model of LIB to analyze the thermal performance and heat generation more accurately. Despite estimation accuracy, the modelling complexity and computational burden limit its application in embedded BMS.

Due to a lack of clear understanding of the electrochemical processes inside the LIB and their corresponding mathematical equations alongside to reduce the computational expenses, often all heat generation sources were not modelled/considered. These unmodeled heat generation behaviors lead to significant errors in temperature estimation. Regarding this, Zhang et al. [67] developed a two-state thermal model utilizing discretization and inverse model techniques which do not require prior knowledge of thermal boundary conditions. Moreover, the model is capable of estimating the total heat generation of a battery cell, thus, thermal modelling of each heat source is not required and abnormal heat generation can also be detected from the estimation results. The effectiveness and robustness of the model were tested for varying thermal boundary conditions and fast charging conditions. While the strategy is designed for self-heating pouch cells, a similar approach could also be adapted for other types of LIBs. Thus, further research is recommended here. A high-fidelity electrochemical model and onboard measurements such as terminal voltage and current were used by Wang et al. [68] to estimate the cell temperature at a wide range of C-rates during the charging/discharging period. They have also used a dual ensemble Kalman filter (DEKF) which incorporates enhanced single-particle dynamics to relate terminal voltage to battery temperature and Li+ concentration. Besides, modelling complexity and high computational cost, the accurate determination of lithium (Li+) concentration is challenging. Therefore, the application of the model in real-life online prediction is questionable. The spatial distribution of internal temperature in LIB was estimated using a pseudo-2D electrochemical model and soft-constrained dual unscented Kalman filter (DUKF) by Marelli and Corno [69]. It is mainly developed to estimate the Li+ concentration and modelling complexity and computational expenses are very high. However, the approach could be extended for temperature estimation. Smith et al. [62] developed a one-dimensional electrochemical, lumped thermal model to explore pulse power limitations and thermal behavior of a LIB pack. The electrochemical thermal modelling-based temperature estimation strategies proposed by different authors are summarized in Table 1 for a quick reference to the readers. In general, the major limitations of any electrochemical model-based strategies are the modelling complexity and high computational cost making these models unsuitable for online prediction and application at low-cost onboard BMS.

Table 1. Summary of electrochemical thermal modelling-based temperature estimation strategies.

Reference	Types of Models	Important Note
Thomas and Newman [60]	One-dimensional electrochemical model	Not used for temperature estimation
Doyle–Fuller–Newmann model [37,61]	Pseudo-two-dimensional (P2D) model	Not used for temperature estimation but several other researchers used
Al Hallaj et al. [53]	A transient one-dimensional thermal model with lumped parameters	Detailed information of electrodes, electrolytes and separator were considered in heat generation model
Fang et al. [64].	Lumped parameter electrochemical-thermal-coupled model	Can estimate one or two average temperatures, performance of individual electrode at various operating temperatures, constant current and pulsing conditions characteristic were considered, experimentally validated
Gu and Wang [41]	Thermal energy generation model, multiphase micro-macroscopic electrochemical model	Temperature-dependent physicochemical properties and thermal behaviors under various charging conditions were considered. Capable of predicting the average cell temperature as well as the temperature distribution inside a cell, volume-averaging technique, numerical simulations
Kumaresan et al. [42]	One-dimensional thermal model	Thermal dependence of various parameters in the model on different discharge profiles was assessed, validated using experimental and simulation results
Kim et al. [65]	Two-dimensional modelling + Finite element method (FEM)	Able to provide temperature distribution based on potential and current density distribution, MATLAB, validated using experimental and simulation results
Gerver et al. [66]	A multi-dimensional electrochemical thermal model	Thermal properties of each cell layer are considered, experimentally validated
Wang et al. [68]	High-fidelity electrochemical model + onboard measurements + dual ensemble Kalman filter (DEKF)	Wide range of C-rates during the charging/discharging period, MATLAB, validated using experimental and simulation results
Marelli and Corno [69]	Pseudo-2D electrochemical model and soft-constrained dual unscented Kalman filter (DUKF)	Can provide information on the spatial distribution of internal temperature, MATLAB Simulation
Smith et al. [62]	A one-dimensional electrochemical lumped thermal model	Adaptive to different drive-cycles, tested and validated with FUDS and HWFET drive cycles, experimentally validated

4.2. Equivalent Electric Circuit Model-Based Temperature Estimation

An equivalent electric circuit model (EECM) represents the thermal dynamics of LIB using electrical system parameters to develop a heat capacitor–resistor-based battery thermal model. Depending on the number of heat capacitors (number of energy storage elements) two types of models, namely, the first-order model and second-order model have been developed so far in the literature. The first-order model consists of one thermal energy storage element whereas a second-order thermal model consists of two heat capacitors, typically, one for the heat capacitance of the core and the other one is for the cell surface [13]. The second-order model can capture more dynamics than the first-order model. The first-order and second-order thermal models of a LIB cell are shown in Figure 3a,b, respectively. In Figure 3, Q represents the heat generation rate, C_c and C_s are the heat capacitance of core and surface, respectively, T_{in} and T_{out} are the temperatures of core and surface of the cell, respectively. T_{amb} is the ambient temperature.

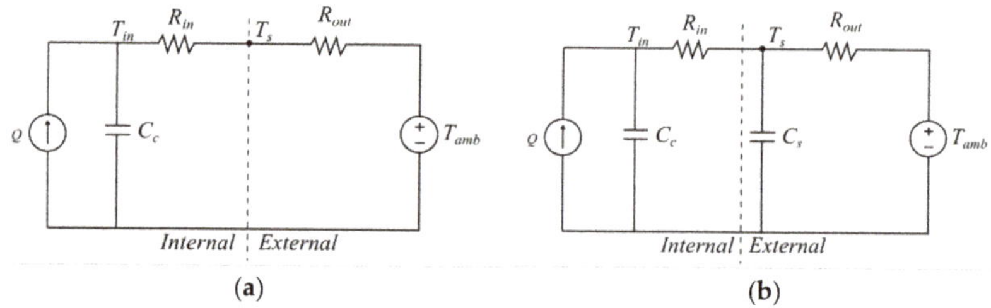

Figure 3. Thermal model of a LIB cell (**a**) First-order model, (**b**) Second-order model.

Further, depending on the modelling complexity, EECM could be also classified as lumped-parameter and distributed parameter models. Lumped-parameter models are used for simplification and thus low computational cost compared to detailed distributed models. Computationally efficient lumped thermal models are developed using single temperature as input to capture the model parameters [70] while some researchers used both surface and core temperatures of the cell to construct the lumped thermal models. Some also considered the correlation between cell geometry and other physical properties with thermal modelling [71]. However, several assumptions were made during modelling leading to inaccurate temperature estimation compared to detailed thermal modelling. Further, thermal models that only estimate the core temperature are considered as single-state/node [72], whereas if the model can estimate both surface and core temperature then it is termed as two-state/node [67] thermal model. The parameters of the EECM are identified through ranges of experimental studies such as electrochemical impedance spectroscopy (EIS) or utilizing externally measurable quantities, such as voltage, current, and temperature. Few studies also considered various conditions of SOC, SOH and estimated surface/core temperatures to make the model more robust. It is very difficult to group those thermal models because lumped models are used in both single-state and dual-state modelling and the model could be first-order and second-order. Therefore, the literature is grouped into cell-level and pack-level temperature estimation schemes that are discussed below.

Typically, these EECM models determine the value of Q using Equation (2) as formulated by Bernardi et al. [20]:

$$Q = I(V - V_{OCV}) + IT_c\frac{dV_{ocv}}{dT_c} \tag{2}$$

where V_{ocv} represents the open-circuit voltage of the battery cell and the term $\frac{dV_{ocv}}{dT_c}$ is the entropy coefficient. Finally, T_c and T_s are estimated using the mathematical form of the thermal models shown in Figure 3. Mathematical equations for temperature estimation using the first-order and second-order thermal model are represented by Equations (3) and (4), respectively.

4.2.1. EECM-Based Cell Temperature Estimation

One of the prime challenges of any EECM-based strategy is model parameter identification. Forgeze et al. [43] used transient experiments by applying current pulses of different magnitudes to increase the internal temperature and the model parameters, heat transfer coefficients and heat capacity were determined to construct a lumped parameter thermal model. This study used EIS for parameter identification where current pulses at 2 Hz were used to increase the internal temperature. The T_c was estimated based on the measured surface temperature using the lumped parameter thermal model. The entropy change was also taken into account while modelling. They developed a first-order thermal model as shown in Figure 3a. The mathematical representation of the first-order thermal model as used by Forgeze et al. reads

$$T_{in} = T_s\left(1 + \frac{R_{in}}{R_{out}}\right) - T_{amb}\frac{R_{in}}{R_{out}} \quad (3)$$

The strategy developed by Forgez et al. lacks quantitative analysis of the influence of heat generation. The operating current is much higher compared to the very low current value used in EIS. Therefore, model parameters determined using EIS are not appropriate for capturing the thermal dynamics accurately. Moreover, they have considered uniform internal temperature, however, more than 10 °C temperature difference among different internal points of a cell has been reported in the same study. This strategy requires surface temperature measurement by installing a temperature sensor at each cell, thus scaling-up is impractical. Maleki and Shamsuri [73] developed a thermal model of notebook computer LIB-pack to understand the thermal response under various operating conditions aiming to reduce the battery pack designing cost and time. They revealed that the temperature rise during charging is dominated by heat dissipation from the control power electronics while during discharging it is dominated by the heat generated inside the LIB cell. These relevant observations must be considered while designing an effective thermal management system of LIB pack, especially for health-conscious fast charging. Surya et al. [13] developed a second-order thermal model for core and surface temperature estimation scheme using KF. Here, the least square (LS) algorithm was employed to identify the battery thermal parameters. Despite the simplicity and good accuracy, environmental uncertainties were not considered during modelling. Moreover, they presented the results based on simulation study alongside very simple and low-current discharge profile was used for model validation, thus, the accuracy in the real-world applications needs further investigation. Previously, models were validated using a simple charging/discharging current profile. However, the load profile in real-life applications much deviates from those simple loading profiles. Therefore, a second-order thermal model and ECM-based two-state thermal model of cylindrical LIB cell were validated with two basic drive-cycle tests, covering an SOC range 25–100%, temperature 5–38 °C, and maximum C-rate of 22 by Lin et al. [74]. The influence of the constantly varying temperature and SOC on the EECM parameters and consequential effect on battery thermal performance was investigated by Lin et al. [74]. The model demonstrated good prediction accuracy and robustness. However, testing using standard internationally referred drive-cycles was not conducted. Thus, accuracy and robustness in practical scenarios need further investigation. EECM parameters are influenced by cell ageing, thus, Li and Yang [75] considered the influences of ageing and heat transfer conditions on thermophysical model parameters. Li and Yang identified the parameters of the extended lumped parameter model online where a forgetting factor recursive least squares (FFRLS) algorithm was employed.

Further to this research, the uncertainties in practical operation were considered by Lin et al. [45,76] alongside the impact of cell ageing during online parameter identification. As an up-gradation, the commonly deployed LS algorithm was augmented with non-uniform forgetting factors to track the time-varying internal parameters making the model adaptive to cell ageing and other uncertainties. In [77], only two lumped models were used to approximate the core and surface temperatures, respectively, which may not be suitable for a large capacity LIB pack due to strong spatiotemporal thermal distribution. While the influence of overpotential entropy changes on battery heat generation was considered, core temperatures estimation of only a single cell was considered. Sun et al. [78] developed a second-order lumped parameter thermal model with the KF technique for core temperature estimation only (single-state). They used an ECM-based heat generation model to mathematically model the accumulation of the total heat generation at the cell core. As an improvement of previous studies, this study considered the influence of entropy changes and overpotential on cell thermal behavior and was quantitively analyzed to develop an online internal temperature estimation strategy. This strategy utilized surface and ambient temperature for core temperature estimation during charge and discharge cycles where the KF was used for adaptive estimation by the process of state and time update in real-time. The impact of unmeasurable modelling error, the initialization error and the possible time-varying external thermal resistance on the temperature estimation accuracy were considered by Dai et al. [79]. In that paper, a second-order lumped parameter thermal model, as shown in Figure 3b, was developed for adaptive core temperature estimation based on the KF. Further, joint Kalman filtering (JKF) was used to simultaneously estimate both core temperature and time-varying external thermal resistance online. The mathematical equation employed by Dai et al. for core temperature estimation can be represented as Equation (4):

$$T_{in}(s) = \frac{\left(1 + \frac{R_{in}}{R_{out}} + C_s R_{in} s\right)}{C_s C_c s^2 + \left(C_s + C_c + \frac{R_{in} C_c}{R_{out}}\right) s + \frac{1}{R_{out}}} Q(s) \tag{4}$$

where s is the Laplace operator. Other parameters are the same as mentioned in Figure 3.

The LS algorithm based on the experimental data was also used to determine the lumped parameters of the thermal model. Dai et al. enhanced the modelling accuracy by constructing a separate thermal model for core and battery shell alongside considering the external heat exchange coefficient as time-varying. The authors simply stated that the proposed method computes efficiently, however, no information about computation time, hardware requirement was presented. Several assumptions were also made during modelling, leading to inaccurate estimation in real-life applications.

A trade-off between the detailed and lumped parameter thermal modelling approaches was considered by Doughty et al. [80] and Park et al. [71]. They developed a two-state thermal model that predicts the surface and core temperature of LIB. The novel intention was to provide more information compared to the lumped model while reducing the computational cost. Few researchers also termed the lumped parameter model as a reduced-order model (ROM). Whilst the primary intention is same, that is, to reduce the complex thermal problem into a simplified heat transfer problem characterized by a reduced set of thermal parameters. A combination of lumped parameter two-state thermal model with 2RC (second-order) ECM along with a joint Kalman filter (JKF)-based core and surface temperature estimation strategy was proposed by Chen et al. [72]. The simulation and experimental test were conducted to verify the adaptiveness of the model to constantly varying temperature and SOC and, finally, the prediction accuracy was also assessed. It was also demonstrated that the proposed model has higher prediction accuracy compared to previously discussed EECMs. It was also demonstrated that the model is highly robust against automatic correction for surface thermal resistance.

To provide more detailed information on the temperature distribution in cylindrical LIB, Xie et al. [81] developed a one-dimensional (radial) lumped parameter thermal model with a dual Kalman filter (DKF). As an improvement, this model is capable to provide temperature information at three different points of the battery, compared to only core and surface temperature. Thus, the researchers termed this modelling as a three-node thermal model. In this study, the anisotropy of thermal conductivity was also considered in identifying internal resistance and SOC during the temperature estimation to enhance the prediction accuracy and robustness. The impact of different charging/discharging current conditions was not considered. Moreover, 1-RC ECM-based heat generation model is considered, thus presumably, the accuracy can be further improved with the application of the 2-RC ECM-based heat generation model. Online parameter estimation using a particle-swarm algorithm with pulse discharge experiments under different ambient temperatures was employed by Pan et al. [19]. A combination of 2RC ECM and a multi-node heat transfer model based on the battery geometry was employed in the study to obtain a more detailed temperature gradient inside the large prismatic LIB. The research showed that the hybrid model could provide similar results to the finite element method (FEM), however, the computational burden was reduced by around 90%. They also revealed that the cell geometry has a strong influence on the cell temperature profile. Despite good accuracy, the effect of cell ageing and the effort of developing pack-level thermal modelling were not considered in this study.

The impact of heat dissipation through radiation from the surface of the cell was introduced in the thermal modelling of LIB by Sun et al. [82]. A lumped thermal model considering the radiation effect was then used for core temperature estimation with the help of an Extended Unscented Kalman Filter (EUKF). The sensor bias was augmented as an extended state to enhance the prediction accuracy and model robustness. While the load profile of residential energy storage was tested, the suitability in commercial vehicle applications was not tested. Further, model parameters were assumed to be constant irrespective of environmental uncertainties which may be in conflict with the facts when the operating conditions will vary significantly. Zhu et al. [83] developed a lumped two-state thermal-electrical model for estimating both the surface and the core temperatures where the thermal impact of the adjacent cell was also considered during modelling. Further, an extended state observer (ESO) with the feedback of the surface temperature was employed to address the model uncertainties and time-variant parameters in the estimation model. This approach is specifically designed for rapid self-heating of self-heating batteries. The concept of model-based virtual thermal sensors (VTS) was introduced by Xiao Y. [84] that combines the tuned thermal model with a KF observer along with an online parameter-identification algorithm for surface and core temperature estimation utilizing a single temperature sensor input. While the strategy is adaptive to environmental uncertainties, it still requires a sensor for feedback; thus, the strategy cannot be termed as completely sensorless. Despite that it minimizes the sensor requirement and enhances the model adaptability, the concept is similar to other lumped parameter EECM-based methods. The effect of fast-discharge on core temperature of LIB was demonstrated by Surya and Mn [14] where a combination of 1-RC ECM, single-state thermal model and KF was used for core temperate estimation. They used a recursive least square (RLS) algorithm to identify model thermal parameters. However, further research is recommended to develop health-conscious BMS suitable for fast charging/discharging.

4.2.2. EECM-Based Temperature Estimation of LIB Pack

Most of the research studies covered only the temperature estimation of a single cell. Thermal modelling and temperature estimation of a LIB pack were seldomly reported. A ROM of a LIB pack considering the characteristic of the inner electrical resistance of the battery was used for core temperature estimation by Ma et al. [85]. Here, RLS was used for the thermal parameter identification. In this study, several assumptions were made while establishing ROM of a battery pack such that parameters of each cell are the same and the thermal behavior of each cell row is same. The heat transfer among cells via conduction through tabs and wires were neglected which could give rise to the error in temperature estimation. Thermal modelling of a LIB pack by scaling-up a single cell thermal model was investigated by Ismail et al. [86] using a simulation study. Considerable accuracy has been noticed, however, several assumptions were made to scale up the single-cell model to battery pack models, such as uniform cell characteristics, constant ambient conditions and 100% efficient discharging process that are far from the real-life scenario. Therefore, the accuracy of the temperature estimation strategy in real-world applications needs to be further explored. Therefore, from the above discussion, it can be stated that the pack-level estimation schemes need significant further research. The EECM-based temperature estimation strategies proposed by different authors are summarized in Table 2 for a quick reference to the readers. One of the major limitations of EECM-based temperature estimation techniques is the requirement of online sensor feedback. This is because the estimation accuracy is completely relying on accuracy of the knowledge of the cell thermal properties, heat generation rates, and thermal boundary conditions represented in terms of electrical parameters that are subjected to change due to cell aging, operating temperature and other practical uncertainties.

Table 2. Summary of EECM-based temperature estimation strategies.

Reference	Types of Models	Important Note
Mahamud et al. [70]	Lumped Parameter heat capacitance–resistance thermal model	ANSYS (Ansys, Inc., Canonsburg, PA, USA) FLUENT Simulation, validated using Experimental and simulation results
Forgeze et al. [43]	Lumped Parameter, Single-State, First-order model	Entropy changes are considered, experimentally validated
Surya et al. [13]	Lumped Parameter, Two-State, Second-order model + Kalman Filter (KF)	SOC, Surface temperature variation, MATLAB Simulation
Lin et al. [74]	Lumped Parameter, Two-State, Second-order model	High current rate, varying temperature, SOC, experimental validation using electrochemical impedance spectroscopy data
Li and Yang [75]	Extended lumped parameter, Two-state, Second-order model + Forgetting factor Recursive Least Square (FFRLS)	Temperature variation, cell ageing, SOC, Heat transfer modes, ANSYS Multiphysics Simulation, validated using experimental and simulation results
Lin et al. [45,76]	Lumped parameter, Two-state model + Least square (LS) algorithm + Nonuniform forgetting factors (NUFF)	Cell ageing and uncertainties in practical operation, validated using experimental and simulation results
Lin et al. [77]	Lumped-parameter model + Closed-loop observer	Influence of overpotential entropy changes, validated using a simulation study
Sun et al. [78]	Lumped parameter, Second-order, Single state thermal model + KF	Influence of entropy changes and overpotential, surface and ambient temperature variation, charge/discharge current profile, MATLAB simulation and experimental validation
Dai et al. [79]	Lumped parameter, Second-order, Two-state model + JKF + LS algorithm	Initialization error and the possible time-varying external thermal resistance, validated using experimental data
Doughty et al. [80] and Park et al. [71]	Lumped parameter, Two-state model + Extended KF	Ambient temperature variation, SOC, validated using a simulation study
Chen et al. [72]	Lumped parameter, Two-state thermal model + Joint KF (JKF)	Constantly varying temperature, SOC, Surface thermal resistance, experimentally validated
Pan et al. [19]	Lumped Parameter, Second-order, multi-node model + particle-swarm algorithm	Battery geometry, charge/discharge profile, Comparison with an FEA model, experimentally validated
Xie et al. [81]	One-dimensional (radial) lumped parameter, Three node model + Dual KF (DKF).	Anisotropy of thermal conductivity, SOC, external temperature, FEM and Computational Fluid Dynamics (CFD), experimental validation
Sun et al. [82]	Lumped parameter, single-state model + Extended unscented KF (EUKF)	Sensor bias, Considered heat radiation from the surface, MATLAB simulation and experimental validation
Zhu et al. [83]	Lumped parameter, Two-state model + extended state observer (ESO)	Thermal impact of an adjacent cell, Model uncertainties and time-variant parameters, MATLAB simulation, validation and comparison using electrochemical impedance spectroscopy data
Surya and Mn [14]	Lumped parameter, Single-state thermal model + KF + Recursive Least Square (RLS) algorithm	Effect of fast-discharge, MATLAB Simulation
Xiao Y. [84]	EECM-based virtual thermal sensors (VTS) + KF	Environmental uncertainties were considered, validated using experimental and simulation results
Ma et al. [85] and Ismail et al. [86]	ROM of a LIB pack for a central temperature of LIB pack + Recursive least square (RLS)	Temperature, SOC, validated using experimental and simulation results

4.3. Numerical Analysis-Based Temperature Estimation

Numerical method-based techniques were successfully implemented for temperature estimation of different chemistries and shapes of LIB cells and even LIB packs. So far, the finite element method (FEM) [87–90] and finite volume method (FVM) [91] were extensively used for temperature estimation. Numerical analysis-based techniques try to mathematically describe the thermal dynamics inside the battery using nonlinear partial differential equations (PDEs) such as used by Du et al. [89]. They have employed FEM analysis with a three-dimensional model and Bernardi equation-based internal heat generation rate. Typically, the PDEs have complex boundary conditions that are infinite-dimensional. The fundamental mathematical equation as employed by Du et al. can be represented as Equation (5)

$$\rho C_p \frac{\partial T}{\partial t} = \lambda_x \frac{\partial^2 T}{\partial x^2} + \lambda_y \frac{\delta^2 T}{\delta y^2} + \lambda_z \frac{\delta^2 T}{\delta z^2} + Q \tag{5}$$

where ρ, C_p represent the mean density and mean specific heat of the cell, respectively. λ is the heat conductivity coefficient of the surface material of the cell and Q is the same as in Equation (1).

Dong Hyup Jeon [87] incorporated a transient thermoelectric model with a porous electrode model and conducted a numerical simulation to understand the thermal behavior of a commercial LIB under charging and discharging conditions. He demonstrated that temperature increase during discharging is much higher compared to the temperature rise during charging. He also suggested that the temperature difference between charge and discharge can be decreased with increasing C-rates. Further, Baba et al. [88] conducted a numerical simulation of an enhanced single-particle model of a LIB to understand the three-dimensional temperature distribution inside the cell. Numerical analysis was used for transient behaviors of a LIB under a dynamic driving cycle by Yi et al. [90]. Double-layer thermal capacitance was used to capture the short-term transient behavior of the LIB chemistry. Fleckenstein et al. [91] using FVM to demonstrate that the temperature gradients inside the cell layer result in different current densities and local SOC inhomogeneities in LIB. These phenomena must be well-taken care of while designing an effective thermal management system. In general, this kind of model is best for capturing both temporally and spatially thermal distribution of the cell as the battery thermal process is a typical distributed parameter system. Despite high accuracy and detailed information about cell temperature gradient, these numerical method-based temperature estimation strategies are not suitable for online temperature estimation due to high computational cost. The complex mathematical analysis also required expertise and strong domain knowledge. Moreover, generalization is not possible as different chemistry and cell physics affect mathematical modelling. A summary of numerical methods-based temperature estimation strategies is shown in Table 3.

Table 3. Summary of numerical methods-based temperature estimation strategies.

Reference	Types of Models	Important Note
Dong Hyup Jeon [87]	A transient thermoelectric model with a porous electrode model + finite element method (FEM)	Different driving cycles, COMSOL Multiphysics (COMSOL Inc., Stockholm, Sweden) simulation and Experimental validation
Baba et al. [88]	Enhanced single-particle model + FEM	Three-dimensional temperature distribution inside the cell, cell geometry, and current profile, experimentally validated
Du et al. [89]	Three-dimensional model + ECM based heat generation model + FEM	Different current profiles, temperature variation, COMSOL Multiphysics simulation and experimental validation

Table 3. *Cont.*

Reference	Types of Models	Important Note
Yi et al. [90]	Transient thermoelectric model + FEM	Transient behaviors under dynamic driving cycle Experimentally validated
Fleckenstein et al. [91]	Three-dimensional model + FVM	Different current density and local SOC inhomogeneities at different cell layers, MATLAB (MathWorks, MA, USA) simulation and experimental validation

4.4. Direct Impedance Measurement-Based Temperature Estimation

Cell internal temperature estimation using a lumped-parameter thermal model and an approximate distributed thermal model have several drawbacks. Firstly, accurate determination of thermal model parameters such as heat generation and cell thermal properties is highly challenging. Heat generation inside the cell is typically approximated by measuring the cell operating current, voltage and the internal resistance that are again functions of SOC, cell internal temperature and SOH. Moreover, a cell is constructed using many different materials combined into a layered structure and thermal contact resistances between these layers are often unknown. Temperature estimation methods use surface temperature measurements and even the combination of surface-mounted temperature sensor and thermal model typically failed to detect the thermal runaway as rapid fluctuations in the internal temperature is difficult to capture using surface mounted sensors because the heat conduction between the core and battery surface takes a considerable amount of time [92]. Furthermore, embedding micro-temperature sensors within the cell [93,94] is not practically possible for a large capacity LIB pack from a manufacturing complexity and system cost point of view. Hence, the core temperature measurement using a physical sensor is not an appropriate method for industrial applications.

Srinivasan et al. [95,96] noticed that the phase of electrochemical impedance in the frequency range of 40 to 100 Hz is temperature-sensitive but insensitive to changes in other parameters such as SOC and SOH. Based on these findings, they demonstrated an electrochemical impedance-based cell internal temperature estimation strategy. However, they assumed the uniform internal temperature and the estimation method is only valid in the temperature range of from -20 to 66 °C. The temperature estimation considering the effect of temperature non-uniformity on electrochemical impedance was studied by Schmidt et al. [97] based on the principle derived by Troxler et al. [98]. Both the strategy developed by Srinivasan et al. and Schmidt et al. were only able to estimate the mean temperature of the cell, however, in real-life application, especially in the case of cylindrical battery under high charging/discharging current, the difference between internal maximum temperature, surface temperature and mean temperature are significantly high. Therefore, Richardson et al. [3] further extended the research and developed a thermal-impedance model by combining an EIS measurement at a single frequency with a surface temperature measurement for precise determination of internal temperature distribution. The fundamental steps in direct impedance measurement-based temperature estimation as presented by Richardson et al. [3] is shown in Figure 4.

The approach of Richardson et al. does not require knowledge of cell thermal properties, heat generation or thermal boundary conditions, however, the major limitation is the online impedance determination of each cell which is highly challenging. Moreover, uncertainties of environmental factors were not considered and a surface-mounted temperature sensor needs to be installed on each cell which is impractical so far. Whilst few approaches of online determination of impedance spectra across multiple frequencies using onboard power electronics of EVs have been reported [99], the application of these strategies in real-time temperature estimation has not yet been investigated. Furthermore, interpreting impedance measurements under superimposed DC currents is yet to be systematically investigated.

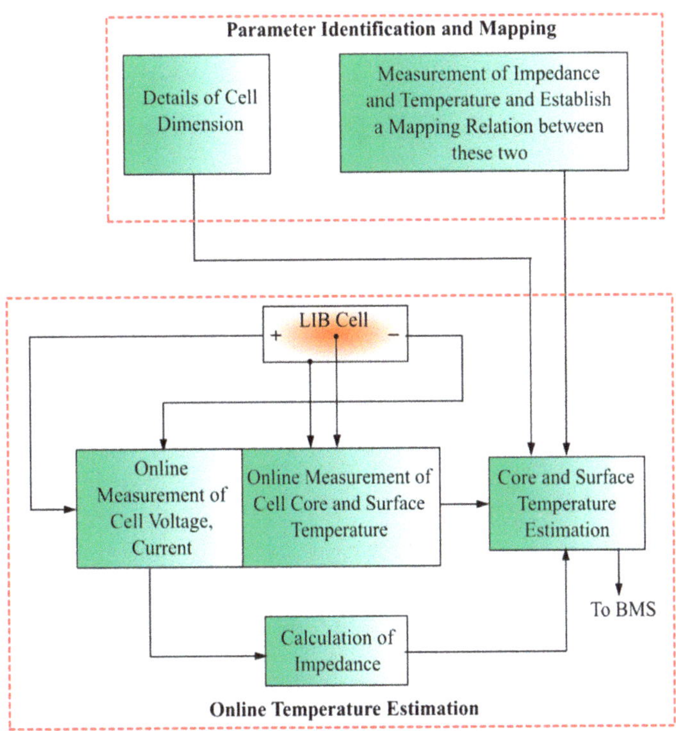

Figure 4. Fundamental steps in direct impedance measurement-based temperature estimation.

Online EIS-based temperature estimation strategy termed impedance-temperature detection (ITD) was proposed by Richardson and Howey [100] for sensorless temperature estimation which is adaptive to cell ageing and practical uncertainties. However, ITD cannot provide a general solution alone, thus, such a strategy combines surface-mounted sensors with ITD for accurate online temperature estimation [3]. Still, temperature sensors are required to be installed. Further to this study, they integrated ITD with an electric-thermal model along with a DEKF for online core temperature estimation of a LIB cell even with unknown convection coefficient. They also demonstrated that the performance of the thermal model plus ITD is almost similar to the ITD with surface thermal sensors. Despite the advantages, the major limitations of the strategy are online impedance determination and the requirement of an accurate electric thermal model, thus encompassing the same drawback of conventional thermal modelling-based strategies. Moreover, although the strategy can estimate both core and surface temperature of an individual cell, the pack-level estimation strategy was not illustrated in this study.

The influence of cell temperature, SOC and SOH on the impedance spectrum, excitation frequency and thereby estimation accuracy of cell internal temperature was investigated by Zhu et al. [101]. Here, the temperature estimation was made based on an impedance response matrix analysis which was developed using EIS measurements. Despite high accuracy, the effect of the nonuniformity of the cell temperature and the correction method was not considered. Moreover, an extensive experimental study is required for modelling and the computational cost is also very high. Thus, the online application of the strategy is challenging. Identification of suitable frequency and other EIS parameters is very difficult whilst the estimation accuracy significantly depends on these parameters. Moreover, accurate determination of the real and imaginary parts of the impedance is highly challenging, whilst different decisions for these two parts leads to inaccurate temperature estimation. A combination of Linear Parameter Varying (LPV) thermal model and a polytopic observer-based battery-cell temperature estimation algorithm was proposed by Debert et al. [102]. The EIS-based strategy was also employed in references [3,103–106] to estimate the core temperature. Despite high accuracy, the major limitation is the determination of accurate impedance-temperature characteristics and it should be acquired in advance through tedious preliminary tests. In addition, the impedance-temperature characteristic of a cell is influenced by cell ageing leading to inaccurate prediction due to SOH deterioration. A summary of direct impedance measurement-based temperature estimation strategies is presented in Table 4.

Table 4. Summary of direct impedance measurement-based strategies.

Reference	Types of Models	Important Note
Srinivasan et al. [95,96]	Direct measurement of electrochemical impedance	Experimental validation with EIS data
Schmidt et al. [97]	Direct measurement of electrochemical impedance	Temperature non-uniformity was not considered, experimentally validated
Richardson et al. [3]	Thermal-impedance model + EIS measurement at single frequency + surface temperature feedback	Independent of cell thermal properties, heat generation or thermal boundary conditions, experimental validation with EIS data
Richardson and Howey [100]	Online EIS measurement (impedance-temperature detection (ITD) + dual-extended Kalman filter (DEKF)	Unknown convection coefficient is considered, experimentally validated
Zhu et al. [101]	Impedance response matrix analysis, developed using EIS measurements	Influence of cell temperature, SOC and SOH on the impedance spectrum, experimental validation with EIS data

4.5. Machine Learning-Based Temperature Estimation

With the overwhelming complexity of the electrochemical reactions inside the battery and the sensitivity of the battery parameters to the uncertainties of the working environment, the thermodynamic behavior varies significantly from the center region to the surface region. Most of the existing distributed thermal models and the lumped parameter thermal models are incapable to consider the spatiotemporal distribution of LIB packs, especially in the case of large-capacity battery packs. Moreover, it is highly difficult to represent these spatiotemporal dynamics by a single physics-based model. Here, the machine learning (ML) algorithms were widely employed to preserve the local dynamics to improve the modelling accuracy of nonlinear systems such as LIB. A schematic layout of the ML-based temperature estimation scheme is shown in Figure 5.

Liu and Li [107] employed a hybrid model of EECM and neural network (NN)-based learning approach to develop a spatiotemporal thermodynamic model of LIB for accurate estimation of internal temperature distribution. The data-driven NN model used commonly measured signals of BMS to compensate for the model-plant mismatch caused by spatial nonlinearity and other model uncertainties. NN and support vector machine (SVM)-based [108] LIB temperature estimation strategy was investigated by Sbarufatti et al. [109]. A hybrid model of an radial basis function neural network (RBFNN) and EKF was employed by Liu et al. [110] to estimate the internal temperature of LIB. While they have considered the impact of temperature on cell behavior, the primary intention of these models was the estimation of SOC or SOH rather than estimating the cell temperature. One of the major challenges of pure ML-based strategies is the generalization capability. Feng et al. [111] developed an effective electrochemical-thermal-neural-network (ETNN) by fusing a lumped parameter electrochemical thermal, feed-forward neural network (FFNN) and a UKF. This method demonstrated appreciable performance in predicting the state of temperature (SOT) in a wide temperature range and large current conditions. However, the modelling is highly complex, the accuracy over different charging current/drive cycles was not tested. Moreover, the computational efficiency and the suitability for online application are questionable. The back of the ETNN is the electrochemical model thus encompassing drawbacks similar to electrochemical models. In general, while ML-based schemes are computationally efficient, collecting training data and model training procedures are highly complex and time expensive. Moreover, real-life battery test data were not considered during ML-based model training in the existing literature; therefore, the accuracy of the existing ML-based strategies is still questionable. A summary of ML-based techniques reported by researchers is presented in Table 5.

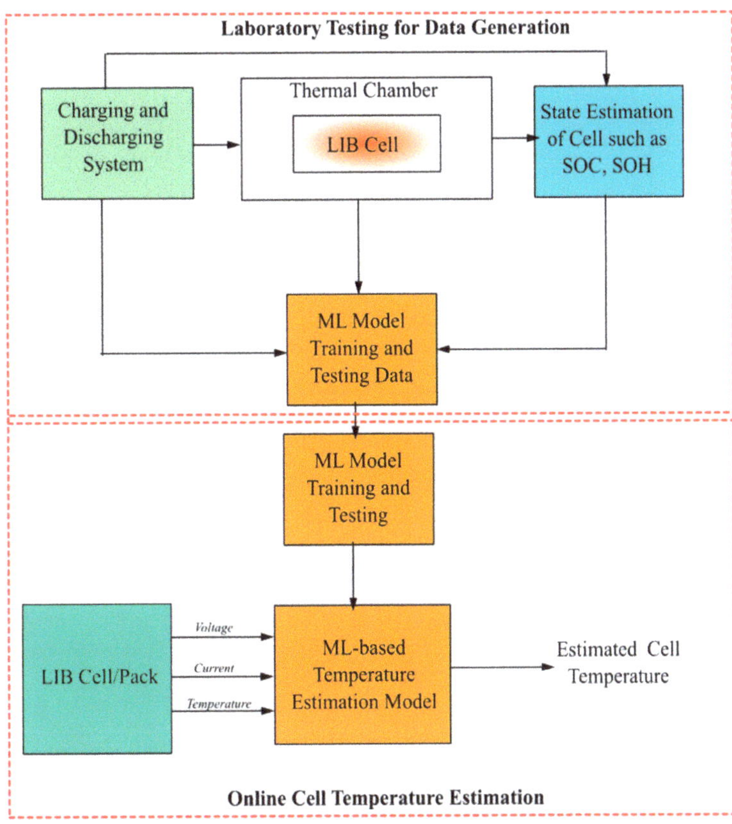

Figure 5. Schematic layout of ML-based temperature estimation scheme.

Table 5. Summary of ML-based temperature estimation techniques.

Reference	Types of Models	Important Note
Liu and Li [107]	EECM + neural network (NN)-based learning approach	Model-plant mismatch caused by spatial nonlinearity and other model uncertainties, NN-model was validated using experimental data
Sbarufatti et al. [109]	Neural networks + Support vector machines	Influence of temperature, charging/discharging current, Python (Python Software Foundation, Wilmington, DE, USA), NN and SVM-model were validated using experimental data

Table 5. *Cont.*

Reference	Types of Models	Important Note
Liu et al. [110]	RBF neural network (RBNN) and the extended Kalman filter (EKF)	Impact of temperature on cell behavior, validated using simulation data
Feng et al. [111]	Electrochemical-thermal-neural-network (ETNN) + Unscented Kalman filter (UKF).	Wide temperature and large current conditions, Python, validated using experimental data

4.6. Magnetic Nanoparticles-Based Temperature Estimation

The magnetization of Magnetic Nanoparticles (MNPs) is nonlinear under an ac magnetic field and the accurate temperature of MNPs could be estimated by using the ratio of the third and fifth harmonic response [112–114]. Further, the temperature sensitivity of MNPs with the increased DC magnetic field was studied by Zhong et al. [115]. They found that the temperature sensitivity of MNPs will decrease with an increased DC magnetic field. Further to this study, Zou et al. [116] developed an improved Magnetic nanoparticles thermometer (MNPT) for the core temperature estimation of LIB which works based on the temperature measurement of magnetic nanoparticles (MNPs). They also suggested the optimal range of the DC magnetic field strength to ensure maximum temperature sensitivity and minimum temperature error of the MNPT. It is noticed that this type of estimation topology is very bulky and costly. Moreover, the suitability of online prediction has not yet been assessed.

5. Discussion on Issues, Challenges and Future Research Recommendations

Temperature estimation schemes for LIBs can be designed with different levels of complexity depending on the requirement of accuracy level and detailing of the prediction results. Detailed model results and more accurate predictions are essential for safer and reliable operation of BMS. However, integrating more detailed cell phenomena into the model eventually increases the modelling complexity, computational cost while, at the same time, reduces the suitability for online prediction and low-cost onboard BMS. For instance, modelling complexity increases if the temperature gradient of each cell layer is considered instead of concentrated heat generation at the core. Secondly, the heat fluxes inside and outside the battery can be considered in both axial and radial directions instead of considering only the radial direction for simplicity. Furthermore, detailed models typically consider different heat transport modes, that is, conduction, convective and radiation whereas simplified models consider only conduction heat transfer. Integrating a greater number of phenomena in thermal modelling requires a lot of parameters, resulting in additional requirements of experimental measurements, modelling time and solid domain knowledge. In addition, very detailed and accurate information of cell structure, material properties and cell assembly are also needed. However, collecting this information from the cell manufacturer is highly challenging due to the confidentiality of the design data. Therefore, it can be inferred from the above discussion that the detailed models could produce highly accurate and complete insight into cell thermodynamics, however, their computational complexity may not be suitable for online prediction and onboard low-cost BMS. In general, most of the estimation strategies require measurements from physical sensors, however, installing a physical sensor at each cell is not practically possible as a high-capacity LIB pack consists of thousands of individual cells. Moreover, installing a sensor at the cell core for core temperature measurement is highly challenging. Several estimation schemes estimate the core temperature based on the surface temperature measurement. However, it is very erroneous as it takes a significant amount of time for the heat to reach the surface from the core. So far, most of the research studies have covered the temperature estimation scheme of a single LIB cell. Temperature estimation of a LIB pack is much more challenging. Thus, significant further research is recommended here. Moreover, the influence of fast charging/discharging on the cell temperature has not yet been deeply explored. It is highly recommended to develop a health-conscious BMS. A summary of existing issues, challenges and future research recommendations to the research community are presented in Table 6.

Table 6. Summary of major issues, challenges and research recommendations.

Strategy	Major Issues and Challenges	Future Research Recommendations
Electrochemical Model-based	• Extremely detailed modelling is possible. Thus, it could produce a highly accurate prediction, however with the expenses of very high computational cost. Thus, unsuitable for online prediction by onboard BMS • In-depth prior knowledge of LIB chemistry is a must besides expertise in mathematical modelling, resulting in dependence on domain experts • Extensive experiments are required to accumulate detailed information on battery characteristics • Modelling is highly complex • Developing an adaptive estimation scheme is highly challenging • Poor generalization capability	• Significant future research is recommended to reduce modelling complexity and computational cost • So far, it can produce the best prediction results, thus could be extensively used for the validation of other types of models and data acquisition for data-driven models • LIB chemistry is highly sensitive to temperature, battery health and other uncertainties, thus, further research on adaptive modelling is recommended
Equivalent Electric Circuit (EECM) Model-based	• Most extensively used so far due to adequate accuracy and easy implementation, however, modelling complexity and computational cost increase with the order of the model, number of temperature measurement points (nodes) and the parameter distribution • Accurate EECM parameters are very difficult to identify, especially online parameter estimation • Parameter tuning using external measurement is challenging and time expensive • Few researchers also used electrochemical analysis for parameter identification and determination which possesses similar difficulties to electrochemical-based strategies • Predictions are highly influenced by measurement noises and often too many physical sensors are required • Lower order models/simplified models are so far extensively used for online prediction with the compromise of accuracy and detailed insight	• Modelling complexity and Computational cost can be controlled by treading-off between accuracy requirement and detailing of the model • Adaptive parameterization is challenging, however with the fusion of advanced algorithms such as ML-based techniques, adaptive strategies could be developed • These models can generate highly accurate results at the laboratory, thus could be used to generate data and model validation of other strategies • Fusion of this strategy with other strategies such as ML-based techniques could produce enhanced accuracy and computational performance • Instead of traditional filters, more advanced adaptive filtering techniques could be embedded for better performance
Machine Learning (ML)-based	• Completely data-driven black-box strategy, that is, prediction depends on the external measurements only, thus, minimal or no requirement of any domain-specific knowledge, however, the major challenge is the accumulation of high-quality large volume of training data • No requirement of iterative complex mathematical calculation, thus, computational cost is adequate for online application, however, computational cost increases with the high volume (high resolution) data and number of feature vectors to obtain a better insight • Accumulation of high-resolution data especially manufacturer data and fault data are highly challenging. These data are important for accurate and adaptive prediction • Generalization is challenging • Currently not used on onboard BMS due to high training time and complex algorithm development and computational time, whilst it is noticed that very few efforts have been made so far • Often, external measurements by physical sensors are required as feedback for online parameter adjustment, thus still requires installation of physical sensors	• While it is comparatively easy to develop adaptive models, however, very few efforts have been made so far. • Cell characteristics are highly influenced by temperature, ageing and other uncertainties, thus, further research on adaptive modelling is recommended • Generalization is difficult, however, with the incorporation of advanced adaptive algorithms, it could be possible • With proper design efforts, it could be used for online prediction and implemented in onboard BMS with low processing power • Very promising technology could be used for a future generation of sensorless temperature estimation strategies. Very little effort has been provided so far, thus, further research is recommended

Table 6. *Cont.*

Strategy	Major Issues and Challenges	Future Research Recommendations
Numerical Model-based	• These strategies use FEA and FVA. FEA and FVA based temperature estimation strategies are considered the most accurate and most computationally expensive • Due to iterative complex mathematical calculation, its computational cost is very high, thus not suitable for online prediction	• Significant research and development are required to improve computational cost to make this suitable for online prediction.As it is most accurate, it could be used for other model validation and accurate data collection
Direct Impedance Measurement-based	• The influence of temperature on cell impedance is used for internal temperature estimation. However, online direct measurement of impedance using onboard power electronics is highly challenging • Changes in cell impedance due to temperature variation is small, thus, accurate determination of such small changes is highly difficult • Existing schemes are very bulky • Very few research efforts have so far provided, not yet practically implemented	• Promising technology, thus, significant further research and development is recommended to reduce scheme size and assess the practical applicability in onboard BMS • Accuracy in real-world applications needs to be judged • Further research into online impedance determination using onboard electronics is also recommended • The cost of existing solutions is very high, which needs to be addressed
Magnetic Nanoparticle-based	• Very new technology, it is too early to comment	• Practical applicability in onboard low-cost BMS has not yet been investigated. Overall, significant further research is required

6. Conclusions

This article presented a comprehensive review of the state-of-the-art temperature estimation strategies for lithium-ion batteries (LIBs) covering the necessity of an optimum estimation strategy, detailed discussion on the existing strategies, current issues, challenges and future research recommendations. It can be inferred that an accurate temperature estimation of LIBs is indispensable for effective thermal management, operational safety and several other crucial tasks of a Battery Management System (BMS). Measurement of each cell temperature using physical sensors is not practically possible, especially for a high-capacity battery pack consisting of thousands of individual cells. To develop an ideal temperature estimation scheme, one needs to concentrate on several factors, such as high accuracy, adaptability, small size, real-time estimation, distribution (to monitor the temperature gradient of the entire cell), low cost, and easily implementable for wide adoption. Typically, a temperature estimation scheme consists of a heat generation model and a heat transfer model. Depending on the modelling and computation strategies temperature estimation schemes can be grouped into six categories, namely, electrochemical model-based, equivalent electric circuit model (EECM)-based, machine learning (ML)-based, numerical analysis-based, direct impedance measurement-based, and magnetic nanoparticle-based. So far, numerical analysis-based schemes are most accurate followed by electro-chemical model-based schemes. However, both strategies have very high computational cost making them inappropriate for online prediction by a low-cost onboard BMS. Moreover, modelling complexity and experimental requirements are very high alongside the necessity of domain-specific knowledge. EECM-based schemes can be designed with different levels of complexity, accuracy level and computational cost. Simplified lower-order EECM-based schemes are extensively used in the literature and practice. Machine learning (ML)-based schemes are very promising due to their higher level of accuracy, ease of implementation and adaptability. In addition, reduced or even no requirement of equivalent modelling and domain experts. However, to obtain the feature vectors, very large volume and high-quality data are required which are typically very challenging to acquire. Here, a hybrid strategy combining an EECM and an ML is presumably a suitable solution. Direct impedance measurement and magnetic nanoparticle-based schemes are very newly developed. It is too early to assess their capability and suitability for online prediction and implementation in onboard BMS. Therefore, systematic guidelines about open research areas and future research directions are highlighted in this study. It is also noticed that the majority of the research studies proposed temperature estimation schemes of a single LIB cell whereas temperature estimation of a LIB pack is much more challenging. Thus, significant further research is recommended here a well.

Author Contributions: Conceptualization, S.S.W.; methodology, A.S. and S.S.W.; investigation, A.S.; resources, S.S.W.; writing—original draft preparation, A.S.; writing—review and editing, A.S. and S.S.W.; visualization, A.S.; supervision, S.S.W. All authors have read and agreed to the published version of the manuscript.

Funding: This research received no external funding.

Institutional Review Board Statement: Not applicable.

Informed Consent Statement: Not applicable.

Data Availability Statement: Not applicable.

Conflicts of Interest: The authors declare no conflict of interest.

References

1. Jilte, R.D.; Kumar, R.; Ma, L. Thermal performance of a novel confined flow Li-ion battery module. *Appl. Therm. Eng.* **2019**, *146*, 1–11. [CrossRef]
2. Yang, X.-H.; Tan, S.-C.; Liu, J. Thermal management of li-ion battery with liquid metal. *Energy Convers. Manag.* **2016**, *117*, 577–585. [CrossRef]
3. Richardson, R.R.; Ireland, P.T.; Howey, D. Battery internal temperature estimation by combined impedance and surface temperature measurement. *J. Power Sources* **2014**, *265*, 254–261. [CrossRef]
4. Wang, Y.-F.; Wu, J.-T. Performance improvement of thermal management system of lithium-ion battery module on purely electric AUVs. *Appl. Therm. Eng.* **2019**, *146*, 74–84. [CrossRef]
5. Surya, S.; Samanta, A.; Marcis, V.; Williamson, S.S. Smart core and surface temperature estimation techniques for health-conscious lithium-ion battery management systems: A model-to-model comparison. *Preprints* **2021**, 1–21. [CrossRef]
6. Tanim, T.R.; Rahn, C.D.; Wang, C.-Y. State of charge estimation of a lithium ion cell based on a temperature dependent and electrolyte enhanced single particle model. *Energy* **2015**, *80*, 731–739. [CrossRef]
7. Farmann, A.; Sauer, D.U. A study on the dependency of the open-circuit voltage on temperature and actual aging state of lithium-ion batteries. *J. Power Sources* **2017**, *347*, 1–13. [CrossRef]
8. Zheng, F.; Jiang, J.; Sun, B.; Zhang, W.; Pecht, M. Temperature dependent power capability estimation of lithium-ion batteries for hybrid electric vehicles. *Energy* **2016**, *113*, 64–75. [CrossRef]
9. Samanta, A.; Chowdhuri, S. Active cell balancing of lithium-ion battery pack using dual dc-dc converter and auxiliary lead-acid battery. *J. Energy Storage* **2021**, *33*, 102109. [CrossRef]
10. Samanta, A.; Chowdhuri, S.; Williamson, S. Machine learning-based data-driven fault detection/diagnosis of lithium-ion battery: A critical review. *Electronics* **2021**, *10*, 1309. [CrossRef]
11. Zhang, S.; Xu, K.; Jow, T. The low temperature performance of li-ion batteries. *J. Power Sources* **2003**, *115*, 137–140. [CrossRef]
12. Mohan, S.; Kim, Y.; Stefanopoulou, A.G. Energy-conscious warm-up of li-ion cells from subzero temperatures. *IEEE Trans. Ind. Electron.* **2016**, *63*, 2954–2964. [CrossRef]
13. Surya, S.; Marcis, V.; Williamson, S. Core temperature estimation for a lithium ion 18650 cell. *Energies* **2020**, *14*, 87. [CrossRef]
14. Surya, S.; Mn, A. Effect of fast discharge of a battery on its core temperature. In Proceedings of the 2020 International Conference on Futuristic Technologies in Control Systems & Renewable Energy (ICFCR), Kerala, India, 23–24 September 2020. [CrossRef]
15. Wei, Z.; Zhao, J.; He, H.; Ding, G.; Cui, H.; Liu, L. Future smart battery and management: Advanced sensing from external to embedded multi-dimensional measurement. *J. Power Sources* **2021**, *489*, 229462. [CrossRef]
16. Steinhorst, S.; Lukasiewycz, M.; Narayanaswamy, S.; Kauer, M.; Chakraborty, S. Smart cells for embedded battery management. In Proceedings of the 2014 IEEE International Conference on Cyber-Physical Systems, Networks, and Applications, Hong Kong, China, 25–26 August 2014.
17. Huang, X.; Sui, X.; Stroe, D.-I.; Teodorescu, R. A review of management architectures and balancing strategies in smart batteries. In Proceedings of the IECON 2019—45th Annual Conference of the IEEE Industrial Electronics Society, Lisbon, Portugal, 14–17 October 2019.
18. Fleming, J.; Amietszajew, T.; Charmet, J.; Roberts, A.J.; Greenwood, D.; Bhagat, R. The design and impact of In Situ and operando thermal sensing for smart energy storage. *J. Energy Storage* **2019**, *22*, 36–43. [CrossRef]
19. Pan, Y.-W.; Hua, Y.; Zhou, S.; He, R.; Zhang, Y.; Yang, S.; Liu, X.; Lian, Y.; Yan, X.; Wu, B. A computational multi-node electro-thermal model for large prismatic lithium-ion batteries. *J. Power Sources* **2020**, *459*, 228070. [CrossRef]
20. Bernardi, D.M.; Pawlikowski, E.M.; Newman, J. A general energy balance for battery systems. *J. Electrochem. Soc.* **1985**, *132*, 5–12. [CrossRef]

21. Sun, F.; Xiong, R.; He, H.; Li, W.; Aussems, J.E.E. Model-based dynamic multi-parameter method for peak power estimation of lithium-ion batteries. *Appl. Energy* **2012**, *96*, 378–386. [CrossRef]
22. Lu, L.; Han, X.; Li, J.; Hua, J.; Ouyang, M. A review on the key issues for lithium-ion battery management in electric vehicles. *J. Power Sources* **2013**, *226*, 272–288. [CrossRef]
23. Ghalkhani, M.; Bahiraei, F.; Nazri, G.-A.; Saif, M. Electrochemical-thermal model of pouch-type lithium-ion batteries. *Electrochim. Acta* **2017**, *247*, 569–587. [CrossRef]
24. Yang, X.-G.; Leng, Y.; Zhang, G.; Ge, S.; Wang, C.-Y. Modeling of lithium plating induced aging of lithium-ion batteries: Transition from linear to nonlinear aging. *J. Power Sources* **2017**, *360*, 28–40. [CrossRef]
25. Chen, M.; Bai, F.; Song, W.; Lv, J.; Lin, S.; Feng, Z.; Li, Y.; Ding, Y. A multilayer electro-thermal model of pouch battery during normal discharge and internal short circuit process. *Appl. Therm. Eng.* **2017**, *120*, 506–516. [CrossRef]
26. Zhao, Y.; Diaz, L.B.; Patel, Y.; Zhang, T.; Offer, G.J. How to cool lithium ion batteries: Optimising cell design using a thermally coupled model. *J. Electrochem. Soc.* **2019**, *166*, A2849–A2859. [CrossRef]
27. Damay, N.; Forgez, C.; Bichat, M.-P.; Friedrich, G. Thermal modeling of large prismatic $LiFePO_4$/graphite battery. Coupled thermal and heat generation models for characterization and simulation. *J. Power Sources* **2017**, *283*, 37–45. [CrossRef]
28. Allafi, W.; Zhang, C.; Uddin, K.; Worwood, D.; Dinh, T.Q.; Ormeno, P.A.; Li, K.; Marco, J. A lumped thermal model of lithium-ion battery cells considering radiative heat transfer. *Appl. Therm. Eng.* **2018**, *143*, 472–481. [CrossRef]
29. Esmaeili, J.; Jannesari, H. Developing heat source term including heat generation at rest condition for lithium-ion battery pack by up scaling information from cell scale. *Energy Convers. Manag.* **2017**, *139*, 194–205. [CrossRef]
30. Arora, S.; Shen, W.; Kapoor, A. Neural network based computational model for estimation of heat generation in $LiFePO_4$ pouch cells of different nominal capacities. *Comput. Chem. Eng.* **2017**, *101*, 81–94. [CrossRef]
31. Hu, X.; Liu, W.; Lin, X.; Xie, Y. A comparative study of control-oriented thermal models for cylindrical li-ion batteries. *IEEE Trans. Transp. Electrif.* **2019**, *5*, 1237–1253. [CrossRef]
32. Xie, Y.; Li, W.; Hu, X.; Zou, C.; Feng, F.; Tang, X. Novel mesoscale electrothermal modeling for lithium-ion batteries. *IEEE Trans. Power Electron.* **2019**, *35*, 2595–2614. [CrossRef]
33. Friesen, A.; Mönnighoff, X.; Börner, M.; Haetge, J.; Schappacher, F.M.; Winter, M. Influence of temperature on the aging behavior of 18650-type lithium ion cells: A comprehensive approach combining electrochemical characterization and post-mortem analysis. *J. Power Sources* **2017**, *342*, 88–97. [CrossRef]
34. Fan, Y.; Bao, Y.; Ling, C.; Chu, Y.; Tan, X.; Yang, S. Experimental study on the thermal management performance of air cooling for high energy density cylindrical lithium-ion batteries. *Appl. Therm. Eng.* **2019**, *155*, 96–109. [CrossRef]
35. Saw, L.H.; Poon, H.M.; Thiam, H.S.; Cai, Z.; Chong, W.T.; Pambudi, N.A.; King, Y.J. Novel thermal management system using mist cooling for lithium-ion battery packs. *Appl. Energy* **2018**, *223*, 146–158. [CrossRef]
36. Liu, B.; Yin, S.; Xu, J. Integrated computation model of lithium-ion battery subject to nail penetration. *Appl. Energy* **2016**, *183*, 278–289. [CrossRef]
37. Doyle, M.; Fuller, T.F.; Newman, J.S. Modeling of galvanostatic charge and discharge of the lithium/polymer/insertion cell. *J. Electrochem. Soc.* **1993**, *140*, 1526–1533. [CrossRef]
38. Xiao, Y.; Fahimi, B. State-space based multi-nodes thermal model for lithium-ion battery. In Proceedings of the 2014 IEEE Transportation Electrification Conference and Expo (ITEC), Beijing, China, 31 August 2014.
39. Tian, N.; Fang, H.; Wang, Y. 3-D Temperature field reconstruction for a lithium-ion battery pack: A distributed kalman filtering approach. *IEEE Trans. Control. Syst. Technol.* **2019**, *27*, 847–854. [CrossRef]
40. Ruan, H.; Jiang, J.; Sun, B.; Gao, W.; Wang, L.; Zhang, W. Online estimation of thermal parameters based on a reduced wide-temperature-range electro-thermal coupled model for lithium-ion batteries. *J. Power Sources* **2018**, *396*, 715–724. [CrossRef]
41. Gu, W.B.; Wang, C.Y. Thermal-electrochemical modeling of battery systems. *J. Electrochem. Soc.* **2000**, *147*, 2910–2922. [CrossRef]
42. Kumaresan, K.; Sikha, G.; White, R.E. Thermal model for a li-ion cell. *J. Electrochem. Soc.* **2008**, *155*, A164–A171. [CrossRef]
43. Forgez, C.; Do, D.V.; Friedrich, G.; Morcrette, M.; Delacourt, C. Thermal modeling of a cylindrical $LiFePO_4$/graphite lithium-ion battery. *J. Power Sources* **2010**, *195*, 2961–2968. [CrossRef]
44. Dees, D.W.; Battaglia, V.S.; Bélanger, A. Electrochemical modeling of lithium polymer batteries. *J. Power Sources* **2002**, *110*, 310–320. [CrossRef]
45. Lin, X.; Perez, H.E.; Siegel, J.; Stefanopoulou, A.G.; Li, Y.; Anderson, R.D.; Ding, Y.; Castanier, M. Online parameterization of lumped thermal dynamics in cylindrical lithium ion batteries for core temperature estimation and health monitoring. *IEEE Trans. Control Syst. Technol.* **2012**, *21*, 1745–1755. [CrossRef]
46. Choi, J.W.; Aurbach, D. Promise and reality of post-lithium-ion batteries with high energy densities. *Nat. Rev. Mater.* **2016**, *1*, 16013. [CrossRef]
47. Yang, F.; Wang, D.; Zhao, Y.; Tsui, K.-L.; Bae, S.J. A study of the relationship between coulombic efficiency and capacity degradation of commercial lithium-ion batteries. *Energy* **2018**, *145*, 486–495. [CrossRef]
48. Wang, Q.; Zhao, X.; Ye, J.; Sun, Q.; Ping, P.; Sun, J. Thermal response of lithium-ion battery during charging and discharging under adiabatic conditions. *J. Therm. Anal. Calorim.* **2016**, *124*, 417–428. [CrossRef]
49. Chen, Y. Heat transfer phenomena in lithium/polymer-electrolyte batteries for electric vehicle application. *J. Electrochem. Soc.* **1993**, *140*, 1833–1838. [CrossRef]

50. Fuller, T.F.; Doyle, M.; Newman, J. Relaxation phenomena in lithium-ion-insertion cells. *J. Electrochem. Soc.* **1994**, *141*, 982–990. [CrossRef]
51. Pals, C.R.; Newman, J. Thermal modeling of the lithium/polymer battery: I discharge behavior of a single cell. *J. Electrochem. Soc.* **1995**, *142*, 3274–3281. [CrossRef]
52. Pals, C.R.; Newman, J. Thermal modeling of the lithium/polymer battery: II temperature profiles in a cell stack. *J. Electrochem. Soc.* **1995**, *142*, 3282–3288. [CrossRef]
53. Al Hallaj, S.; Maleki, H.; Hong, J.; Selman, J. Thermal modeling and design considerations of lithium-ion batteries. *J. Power Sources* **1999**, *83*, 1–8. [CrossRef]
54. Gomadam, P.M.; White, R.E.; Weidner, J.W. Modeling heat conduction in spiral geometries. *J. Electrochem. Soc.* **2003**, *150*, A1339–A1345. [CrossRef]
55. Chen, Y.; Evans, J.W. Three-dimensional thermal modeling of lithium-polymer batteries under galvanostatic discharge and dynamic power profile. *J. Electrochem. Soc.* **1994**, *141*, 2947–2955. [CrossRef]
56. Newman, J.; Tiedemann, W. Temperature rise in a battery module with constant heat generation. *J. Electrochem. Soc.* **1995**, *142*, 1054–1057. [CrossRef]
57. Chen, Y.; Evans, J.W. Thermal analysis of lithium-ion batteries. *J. Electrochem. Soc.* **1996**, *143*, 2708–2712. [CrossRef]
58. Rao, L.; Newman, J. Heat-generation rate and general energy balance for insertion battery systems. *J. Electrochem. Soc.* **1997**, *144*, 2697–2704. [CrossRef]
59. Kim, G.-H.; Pesaran, A.; Spotnitz, R. A three-dimensional thermal abuse model for lithium-ion cells. *J. Power Sources* **2007**, *170*, 476–489. [CrossRef]
60. Thomas, K.E.; Newman, J.; Darling, R.M. Mathematical modeling of lithium batteries. In *Advances in Lithium-Ion Batteries*; Springer: Berlin/Heidelberg, Germany, 2002; pp. 345–392.
61. Van Schalkwijk, W.; Scrosati, B. Advances in lithium ion batteries introduction. In *Advances in Lithium-Ion Batteries*; Springer: Berlin/Heidelberg, Germany, 2002.
62. Smith, K.; Wang, C.-Y. Power and thermal characterization of a lithium-ion battery pack for hybrid-electric vehicles. *J. Power Sources* **2006**, *160*, 662–673. [CrossRef]
63. Khateeb, S.A.; Amiruddin, S.; Farid, M.; Selman, J.R.; Al-Hallaj, S. Thermal management of li-ion battery with phase change material for electric scooters: Experimental validation. *J. Power Sources* **2005**, *142*, 345–353. [CrossRef]
64. Fang, W.; Kwon, O.J.; Wang, C.-Y. Electrochemical-thermal modeling of automotive li-ion batteries and experimental validation using a three-electrode cell. *Int. J. Energy Res.* **2009**, *34*, 107–115. [CrossRef]
65. Kim, U.S.; Yi, J.; Shin, C.B.; Han, T.; Park, S. Modelling the thermal behaviour of a lithium-ion battery during charge. *J. Power Sources* **2011**, *196*, 5115–5121. [CrossRef]
66. Gerver, R.E.; Meyers, J.P. Three-dimensional modeling of electrochemical performance and heat generation of lithium-ion batteries in tabbed planar configurations. *J. Electrochem. Soc.* **2011**, *158*, A835–A843. [CrossRef]
67. Zhang, J.; Yang, X.-G.; Sun, F.; Wang, Z.; Wang, C.-Y. An online heat generation estimation method for lithium-ion batteries using dual-temperature measurements. *Appl. Energy* **2020**, *272*, 115262. [CrossRef]
68. Wang, P.; Yang, L.; Wang, H.; Tartakovsky, D.; Onori, S. Temperature estimation from current and voltage measurements in lithium-ion battery systems. *J. Energy Storage* **2021**, *34*, 102133. [CrossRef]
69. Marelli, S.; Corno, M. Model-based estimation of lithium concentrations and temperature in batteries using soft-constrained dual unscented kalman filtering. *IEEE Trans. Control. Syst. Technol.* **2021**, *29*, 926–933. [CrossRef]
70. Mahamud, R.; Park, C. Reciprocating air flow for li-ion battery thermal management to improve temperature uniformity. *J. Power Sources* **2011**, *196*, 5685–5696. [CrossRef]
71. Park, C.; Jaura, A.K. *Dynamic Thermal Model of Li-Ion Battery for Predictive Behavior in Hybrid and Fuel Cell Vehicles*; SAE Technical Paper Series; SAE International: Warrendale, PA, USA, 2003.
72. Chen, L.; Hu, M.; Cao, K.; Li, S.; Su, Z.; Jin, G.; Fu, C. Core temperature estimation based on electro-thermal model of lithium-ion batteries. *Int. J. Energy Res.* **2020**, *44*, 5320–5333. [CrossRef]
73. Maleki, H.; Shamsuri, A.K. Thermal analysis and modeling of a notebook computer battery. *J. Power Sources* **2003**, *115*, 131–136. [CrossRef]
74. Lin, X.; Perez, H.E.; Mohan, S.; Siegel, J.; Stefanopoulou, A.G.; Ding, Y.; Castanier, M.P. A lumped-parameter electro-thermal model for cylindrical batteries. *J. Power Sources* **2014**, *257*, 1–11. [CrossRef]
75. Li, D.; Yang, L. Identification of spatial temperature gradient in large format lithium battery using a multilayer thermal model. *Int. J. Energy Res.* **2019**, *44*, 282–297. [CrossRef]
76. Lin, X.; Stefanopoulou, A.G.; Perez, H.E.; Siegel, J.B.; Li, Y.; Anderson, R.D. Quadruple adaptive observer of the core temperature in cylindrical li-ion batteries and their health monitoring. In Proceedings of the 2012 American Control Conference (ACC), Montreal, QC, Canada, 27–29 June 2012. [CrossRef]
77. Lin, X.; Fu, H.; Perez, H.E.; Siege, J.B.; Stefanopoulou, A.G.; Ding, Y.; Castanier, M.P. Parameterization and observability analysis of scalable battery clusters for onboard thermal management. *Oil Gas Sci. Technol.* **2013**, *68*, 165–178. [CrossRef]
78. Sun, J.; Wei, G.; Pei, L.; Lu, R.; Song, K.; Wu, C.; Zhu, C. Online internal temperature estimation for lithium-ion batteries based on kalman filter. *Energies* **2015**, *8*, 4400–4415. [CrossRef]

79. Dai, H.; Zhu, L.; Zhu, J.; Wei, X.; Sun, Z. Adaptive kalman filtering based internal temperature estimation with an equivalent electrical network thermal model for hard-cased batteries. *J. Power Sources* **2015**, *293*, 351–365. [CrossRef]
80. Doughty, D.H.; Butler, P.C.; Jungst, R.G.; Roth, E. Lithium battery thermal models. *J. Power Sources* **2002**, *110*, 357–363. [CrossRef]
81. Xie, Y.; Li, W.; Hu, X.; Lin, X.; Zhang, Y.; Dan, D.; Feng, F.; Liu, B.; Li, K. An enhanced online temperature estimation for lithium-ion batteries. *IEEE Trans. Transp. Electrif.* **2020**, *6*, 375–390. [CrossRef]
82. Sun, L.; Sun, W.; You, F. Core temperature modelling and monitoring of lithium-ion battery in the presence of sensor bias. *Appl. Energy* **2020**, *271*, 115243. [CrossRef]
83. Zhu, C.; Shang, Y.; Lu, F.; Jiang, Y.; Cheng, C.; Mi, C. Core temperature estimation for self-heating automotive lithium-ion batteries in cold climates. *IEEE Trans. Ind. Inform.* **2019**, *16*, 3366–3375. [CrossRef]
84. Xiao, Y. Model-based virtual thermal sensors for lithium-ion battery in EV applications. *IEEE Trans. Ind. Electron.* **2014**, *62*, 3112–3122. [CrossRef]
85. Ma, Y.; Cui, Y.; Mou, H.; Gao, J.; Chen, H. Core temperature estimation of lithium-ion battery for EVs using kalman filter. *Appl. Therm. Eng.* **2020**, *168*, 114816. [CrossRef]
86. Ismail, N.H.F.; Toha, S.F.; Azubir, N.A.M.; Ishak, N.H.M.; Hassan, I.D.M.K.; Ibrahim, B.S.K. Simplified heat generation model for lithium ion battery used in electric vehicle. *IOP Conf. Ser. Mater. Sci. Eng.* **2013**, *53*, 012014. [CrossRef]
87. Jeon, D.H. Numerical modeling of lithium ion battery for predicting thermal behavior in a cylindrical cell. *Curr. Appl. Phys.* **2014**, *14*, 196–205. [CrossRef]
88. Baba, N.; Yoshida, H.; Nagaoka, M.; Okuda, C.; Kawauchi, S. Numerical simulation of thermal behavior of lithium-ion secondary batteries using the enhanced single particle model. *J. Power Sources* **2014**, *252*, 214–228. [CrossRef]
89. Du, S.; Jia, M.; Cheng, Y.; Tang, Y.; Zhang, H.; Ai, L.; Zhang, K.; Lai, Y. Study on the thermal behaviors of power lithium iron phosphate (LFP) aluminum-laminated battery with different tab configurations. *Int. J. Therm. Sci.* **2015**, *89*, 327–336. [CrossRef]
90. Yi, J.; Lee, J.; Shin, C.B.; Han, T.; Park, S. Modeling of the transient behaviors of a lithium-ion battery during dynamic cycling. *J. Power Sources* **2015**, *277*, 379–386. [CrossRef]
91. Fleckenstein, M.; Bohlen, O.; Roscher, M.A.; Bäker, B. Current density and state of charge inhomogeneities in li-ion battery cells with $LiFePO_4$ as cathode material due to temperature gradients. *J. Power Sources* **2011**, *196*, 4769–4778. [CrossRef]
92. Santhanagopalan, S.; Ramadass, P.; Zhang, J.Z. Analysis of internal short-circuit in a lithium ion cell. *J. Power Sources* **2009**, *194*, 550–557. [CrossRef]
93. Li, Z.; Zhang, J.; Wu, B.; Huang, J.; Nie, Z.; Sun, Y.; An, F.; Wu, N. Examining temporal and spatial variations of internal temperature in large-format laminated battery with embedded thermocouples. *J. Power Sources* **2013**, *241*, 536–553. [CrossRef]
94. Mutyala, M.S.K.; Zhao, J.; Li, J.; Pan, H.; Yuan, C.; Li, X. In Situ temperature measurement in lithium ion battery by transferable flexible thin film thermocouples. *J. Power Sources* **2014**, *260*, 43–49. [CrossRef]
95. Srinivasan, R.; Carkhuff, B.G.; Butler, M.H.; Baisden, A.C. Instantaneous measurement of the internal temperature in lithium-ion rechargeable cells. *Electrochim. Acta* **2011**, *56*, 6198–6204. [CrossRef]
96. Srinivasan, R. Monitoring dynamic thermal behavior of the carbon anode in a lithium-ion cell using a four-probe technique. *J. Power Sources* **2012**, *198*, 351–358. [CrossRef]
97. Schmidt, J.P.; Arnold, S.; Loges, A.; Werner, D.; Wetzel, T.; Ivers-Tiffée, E. Measurement of the internal cell temperature via. impedance: Evaluation and application of a new method. *J. Power Sources* **2013**, *243*, 110–117. [CrossRef]
98. Troxler, Y.; Wu, B.; Marinescu, M.; Yufit, V.; Patel, Y.; Marquis, A.J.; Brandon, N.P.; Offer, G. The effect of thermal gradients on the performance of lithium-ion batteries. *J. Power Sources* **2014**, *247*, 1018–1025. [CrossRef]
99. Brandon, N.; Mitcheson, P.; Yufit, V.; Howey, D.; Offer, G. Battery Monitoring in Electric Vehicles, Hybrid Electric Vehicles and Other Applications. U.S. Patent 2,013,022,915,6A1, 8 May 2013.
100. Richardson, R.R.; Howey, D.A. Sensorless battery internal temperature estimation using a kalman filter with impedance measurement. *IEEE Trans. Sustain. Energy* **2015**, *6*, 1190–1199. [CrossRef]
101. Zhu, J.; Sun, Z.; Wei, X.; Dai, H. A new lithium-ion battery internal temperature on-line estimate method based on electrochemical impedance spectroscopy measurement. *J. Power Sources* **2015**, *274*, 990–1004. [CrossRef]
102. Debert, M.; Colin, G.; Bloch, G.; Chamaillard, Y. An observer looks at the cell temperature in automotive battery packs. *Control Eng. Pract.* **2013**, *21*, 1035–1042. [CrossRef]
103. Hu, X.; Yuan, H.; Zou, C.; Li, Z.; Zhang, L. Co-estimation of state of charge and state of health for lithium-ion batteries based on fractional-order calculus. *IEEE Trans. Veh. Technol.* **2018**, *67*, 10319–10329. [CrossRef]
104. Wang, Z.; Ma, J.; Zhang, L. State-of-health estimation for lithium-ion batteries based on the multi-island genetic algorithm and the gaussian process regression. *IEEE Access* **2017**, *5*, 21286–21295. [CrossRef]
105. Hande, A. Internal battery temperature estimation using series battery resistance measurements during cold temperatures. *J. Power Sources* **2006**, *158*, 1039–1046. [CrossRef]
106. Howey, D.; Mitcheson, P.D.; Yufit, V.; Offer, G.; Brandon, N.P. Online measurement of battery impedance using motor controller excitation. *IEEE Trans. Veh. Technol.* **2014**, *63*, 2557–2566. [CrossRef]
107. Liu, Z.; Li, H. A Spatiotemporal estimation method for temperature distribution in lithium-ion batteries. *IEEE Trans. Ind. Inform.* **2014**, *10*, 2300–2307. [CrossRef]
108. Khelif, R.; Chebel-Morello, B.; Malinowski, S.; Laajili, E.; Fnaiech, F.; Zerhouni, N. Direct remaining useful life estimation based on support vector regression. *IEEE Trans. Ind. Electron.* **2017**, *64*, 2276–2285. [CrossRef]

109. Sbarufatti, C.; Corbetta, M.; Giglio, M.; Cadini, F. Adaptive prognosis of lithium-ion batteries based on the combination of particle filters and radial basis function neural networks. *J. Power Sources* **2017**, *344*, 128–140. [CrossRef]
110. Liu, K.; Li, K.; Peng, Q.; Guo, Y.; Zhang, L. Data-driven hybrid internal temperature estimation approach for battery thermal management. *Complexity* **2018**, *2018*, 1–15. [CrossRef]
111. Feng, F.; Teng, S.; Liu, K.; Xie, J.; Xie, Y.; Liu, B.; Li, K. Co-estimation of lithium-ion battery state of charge and state of temperature based on a hybrid electrochemical-thermal-neural-network model. *J. Power Sources* **2020**, *455*, 227935. [CrossRef]
112. Li, Y.; Liu, W.; Zhong, J. Comparison of noninvasive and remote temperature estimation employing magnetic nanoparticles in DC and AC applied fields. In Proceedings of the 2012 IEEE International Instrumentation and Measurement Technology Conference, Graz, Austria, 13–16 May 2012.
113. Zhong, J.; Dieckhoff, J.; Schilling, M.; Ludwig, F. Influence of static magnetic field strength on the temperature resolution of a magnetic nanoparticle thermometer. *J. Appl. Phys.* **2016**, *120*, 143902. [CrossRef]
114. Yoshida, T.; Enpuku, K. Simulation and quantitative clarification of ac susceptibility of magnetic fluid in nonlinear brownian relaxation region. *Jpn. J. Appl. Phys.* **2009**, *48*, 127002. [CrossRef]
115. Zhong, J.; Liu, W.; Du, Z.; De Morais, P.C.; Xiang, Q.; Xie, Q. A noninvasive, remote and precise method for temperature and concentration estimation using magnetic nanoparticles. *Nanotechnology* **2012**, *23*, 075703. [CrossRef]
116. Zou, D.; Li, M.; Wang, D.; Li, N.; Su, R.; Zhang, P.; Gan, Y.; Cheng, J. Temperature estimation of lithium-ion battery based on an improved magnetic nanoparticle thermometer. *IEEE Access* **2020**, *8*, 135491–135498. [CrossRef]

Article

All-SiC ANPC Submodule for an Advanced 1.5 kV EV Charging System under Various Modulation Methods

Rafał Kopacz, Michał Harasimczuk, Bartosz Lasek, Rafał Miśkiewicz and Jacek Rąbkowski *

Institute of Control and Industrial Electronics, Warsaw University of Technology, 00-662 Warsaw, Poland; rafal.kopacz@pw.edu.pl (R.K.); michal.harasimczuk@pw.edu.pl (M.H.); bartosz.lasek@pw.edu.pl (B.L.); rafal.miskiewicz@pw.edu.pl (R.M.)
* Correspondence: jacek.rabkowski@pw.edu.pl

Abstract: This work is focused on the design and experimental validation of the all-SiC active neutral-point clamped (ANPC) submodule for an advanced electric vehicle (EV) charging station. The topology of the station is based on a three-wire bipolar DC bus (±750 V) connecting an ac grid converter, isolated DC-DC converters, and a non-isolated DC-DC converter with a battery energy storage. Thus, in all types of power converters, the same three-level submodule may be applied. In this paper, a submodule rated at 1/3 of the nominal power of the grid converter (20 kVA) is discussed. In particular, four different modulation strategies for the 1.5 kV ANPC submodule, exclusively employing fast silicon carbide (SiC) MOSFETs, are considered, and their impact on the submodule performance is analyzed. Moreover, the simulation study is included. Finally, the laboratory prototype is described and experimentally verified at a switching frequency of 64 kHz. It is shown that the system can operate with all of the modulations, while techniques PWM2 and PWM3 emerge as the most efficient, and alternating between them, depending on the load, should be considered to maximize the efficiency. Furthermore, the results showcase that the impact of the different PWM techniques on switching oscillations, including overvoltages, can be nearly fully omitted for a parasitic inductance optimized circuit, and the choice of modulation should be based on power loss and/or other factors.

Keywords: ANPC converter; EV charging; multilevel converter; PWM methods; SiC MOSFETs

Citation: Kopacz, R.; Harasimczuk, M.; Lasek, B.; Miśkiewicz, R.; Rąbkowski, J. All-SiC ANPC Submodule for an Advanced 1.5 kV EV Charging System under Various Modulation Methods. *Energies* **2021**, *14*, 5580. https://doi.org/10.3390/en14175580

Academic Editors: Andrei Blinov, Sheldon Williamson and Hugo Morais

Received: 11 July 2021
Accepted: 30 August 2021
Published: 6 September 2021

1. Introduction

There is no doubt that easily available fast charging infrastructure is a necessary condition in the further expansion of electric vehicles (EVs) beyond current numbers, even in the most developed countries [1,2]. In comparison to traditional cars, the charging time of EVs is, and will be, longer than refueling a tank with gasoline. However, fast charging stations may offer a reduction in time from the range of hours to tens of minutes [3]. This is associated with an increase in charging power to hundreds of kWs, and, unfortunately, a rising number of such stations is challenging for the power system. Therefore, an answer to this problem may be a battery energy storage, reducing power peaks during fast charging periods [4,5]. Additionally, the storage may also act as local energy storage for a PV plant, and perform short-time grid support services. All in all, the EV charging station with energy storage becomes a high-power and complex power electronics system, as can be seen in Figure 1. Moreover, to decrease current levels and conduction losses in a common DC-link (750–800 V), a bipolar topology may also be taken into account [6]. All three types of power converters in such a system: grid-connected AC-DC, isolated DC-DC, and non-isolated DC-DC, should be based on one of the multilevel topologies. Furthermore, in the optimal scenario, all of them should be based on the same topology to reduce the complexity and cost of the whole system, and to introduce modularity into the system. While the isolation stage of the AC-side is omitted in this paper, as the focus is on the presented ANPC submodule, it is worth noting that either conventional low-frequency transformers or solid-state transformers [7] are applicable in the EV charging system.

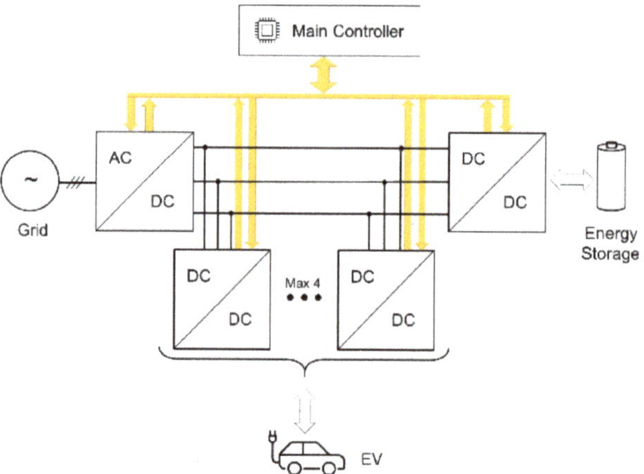

Figure 1. Bipolar DC grid-based EV charging station with energy storage.

Several multilevel converter topologies are applicable in EV charging systems [8]. Most notably, neutral point clamped (NPC), T-type, flying capacitor (FC), and modular multilevel converter (MMC) topologies can be named. In the discussed case, the active neutral point clamped (ANPC) topology of the submodule, introduced in [9], was selected as it is characterized by more flexible control and the possibility to reach a more balanced power loss distribution among the semiconductors compared to the conventional NPC topology, while keeping the power device stress on a similar level [10,11]. Moreover, when compared to other common multilevel structure, such as T-type and FC converter topologies, the former is less efficient for the systems in such voltage range (1500 V) than the NPC-derived topologies [12], whereas the latter cannot be employed in a system with a bipolar DC bus for obvious reasons. Finally, when a comparison with MMC-based systems [13] is considered, ANPC converters seem more appropriate, as MMC introduces a bulkier structure, and adds further complexity to the system [8]. Furthermore, ANPC systems were positively verified in terms of both three-wire [14] and EV charging systems [15].

Moreover, as the SiC technology is constantly developing and providing power semiconductor devices with superior performance compared to conventional Si counterparts [16–18], especially for this specific voltage range, applying SiC MOSFETs to the switches lead to, amongst many, increased efficiency and higher power density. However, briefly after the commercial introduction of SiC power devices, when the SiC technology was still relatively new, the cost of such power devices was significantly higher compared to conventional Si IGBTs. Thus, different hybrid Si/SiC ANPC topologies were introduced, in which SiC MOSFETs can be applied as the power devices for two or four out of six devices per leg depending on the preferred PWM technique, showcasing satisfactory results in terms of achieving a balance between cost and performance [18–22]. Nevertheless, even though the hybrid topologies should be considered when the cost is taken into account, the all-SiC system is still unmatched when strictly performance and maximization of efficiency are considered [19]. Moreover, such a converter structure is and will become even more compelling, as the wide-band-gap technology is currently becoming more and more advanced, and thus SiC power devices will develop to be affordable to a greater extent yearly.

Since the number of switches in the ANPC leg is high, so is the variety of PWM methods applicable in such a system. Moreover, depending on the focus of a specific application, there are modulation techniques that can target various factors, such as efficiency or power density, ensuring equal loss distribution, or, finally lowering line filter requirements [10,23–25]. Furthermore, according to the literature, the impact of commutation path lengths is a crucial matter, determining the proper PWM method for a specific application as well. This is especially relevant when systems with SiC power devices are considered, as wide-band-gap semiconductors are capable of high-speed switching, and thus are more prone to ringing and overvoltages compared to its Si counterparts.

However, in this paper, except for validating the constructed low-volume prototype of the all-SiC ANPC single leg rated at 1500 V DC and 6.67 kVA power (1/3 of three-phase 20 kVA system), it is shown that when enough care and focus is put into the design process of the converter and thus the commutation path lengths are vastly minimized, the variances between different modulation techniques in this regard are not as apparent and the choice may be limited to other factors, namely in this case, efficiency. The conclusions are based on a parasitic inductance optimized ANPC leg that can be used as a submodule to construct full power electronic systems, e.g., three-phase bidirectional AC/DC converters as shown in Figure 1.

Furthermore, in this paper, the PWM techniques are compared based on efficiency and switching performance, strictly for an all-SiC system. In contrast, other researchers have focused on a comparison between different Si/SiC configurations with strictly bound modulation techniques, where each configuration was tested with its specific PWM method. Finally, the conclusion is drawn that while all modulation techniques are viable, two emerge as the most competent, one for lower power ratings and another for higher power ratings. Thus, the assumption is made that to operate optimally, the modulation technique should be changed according to the load. Moreover, while systems comprised of ANPC submodules have been shown in the past, they are connected with other power semiconductor device types, such as IGBTs [26,27] or IGCT [28]. There are no publications regarding SiC MOSFET-based systems rated at MV level, whereas for such an application, the impact of parasitic inductances due to high dv/dt rates and a high switching speed is much more severe, and thus also more critical during the design process [29].

The paper is organized as follows. After the introduction, in Section 2, the basic principles of the ANPC topology are explained together with the considered PWM methods and their operation principles. Then, the simulation study is shown in Section 3, and, in Section 4, the experimental model of the SiC-based submodule is presented along with the results showcasing the experimental validation and further the discussion. Finally, the paper is concluded with a summary in Section 5.

2. Modulation Strategies in Active Neutral Point Clamped (Anpc) Converter

The most popular inverter topology used in industrial power electronic is a basic three-phase two-level (2-L) inverter [30]. This is mainly due to its simple design and well-understood operation principles. However, the voltage stress of semiconductor power devices in such topology is greater than in the DC-link voltage bus. This prevents the use of 1.2 kV SiC power devices in 2-L inverters with greater DC voltage. A well-known alternative to 2-L inverters are three-level (3-L) inverters [31]. The use of such a topology provides a halved maximum voltage stress in semiconductor power devices. Furthermore, due to the three-level nature of the output filter inductor voltage in 3-L inverters, it is possible to reduce the harmonic distortion and output filter volume [32,33]. The low switching time of SiC power MOSFETs also allows a reduction in switching losses. On the other hand, the high value of stray inductance and di/dt of transistors during commutation lead to voltage overshoots, which have a negative impact on MOSFETs' lifetime, energy conversion efficiency, and EMI. In 3-L inverters' three different operation states (positive state P, zero state 0, and negative state N) can be recognized, differing in voltage applied to the inductor. A highly regarded 3-L inverter topology is the ANPC, used in the discussed submodule. The ANPC converter consists of six active switches, S_1–S_6, connected according to Figure 2.

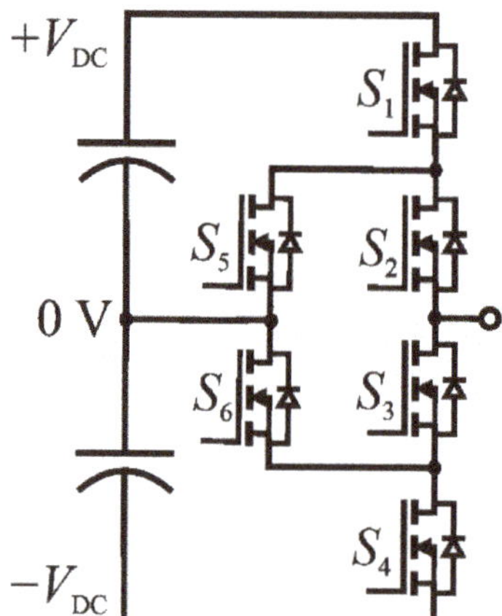

Figure 2. A single-phase leg of the ANPC converter with SiC MOSFETs.

In the ANPC converter, when the output voltage is positive, the inverter is switching between positive $+V_{DC}$ voltage and zero voltage; and when the output voltage is negative, the inverter is switching between negative $-V_{DC}$ and zero voltage. Control of the converter when the voltage is positive and negative is analogous. Therefore, in this article, the different control methods are only described when the voltage is positive. P and N states can be obtained only by turning on transistors S_1 and S_2 in P state, and S_3 and S_4 in N state. During P state, transistor S_6 can be turned on. Similarly, during N state, transistor S_5 can be on as well. This ensures constant v_{DS} voltages equal to V_{DC} on transistors S_3 and S_4 during P state, and on S_1 and S_2 during N state. Simultaneously, in the ANPC topology, there are different approaches to obtain zero state. Four modulation strategies of the ANPC and one of the NPC converters are depicted in Table 1 and Figure 3. In this paper, there are four different modulations described (PWM1–PWM4). These PWM techniques differ from each other in regard to the zero state, in which the current flows through different conduction paths marked in Figure 4 by 2 and 3.

Table 1. Switching states of the ANPC inverter.

State	S_1	S_2	S_3	S_4	S_5	S_6	Conduction Path(s)	PWM Method
P	1	1	0	0	0	1/0	1	1, 2, 3, 4
0U3	1	0	1	0	0	1	3	2
0U2	0	1	0	0	1	0	2	1
0U1	0	1	0	1	1	0	2	4
0F	0	1	1	0	1	1	2 and 3	3
0L1	1	0	1	0	0	1	3	4
0L2	0	0	1	0	0	1	3	1
0L3	0	1	0	1	1	0	2	2
N	0	0	1	1	1/0	0	4	1, 2, 3, 4

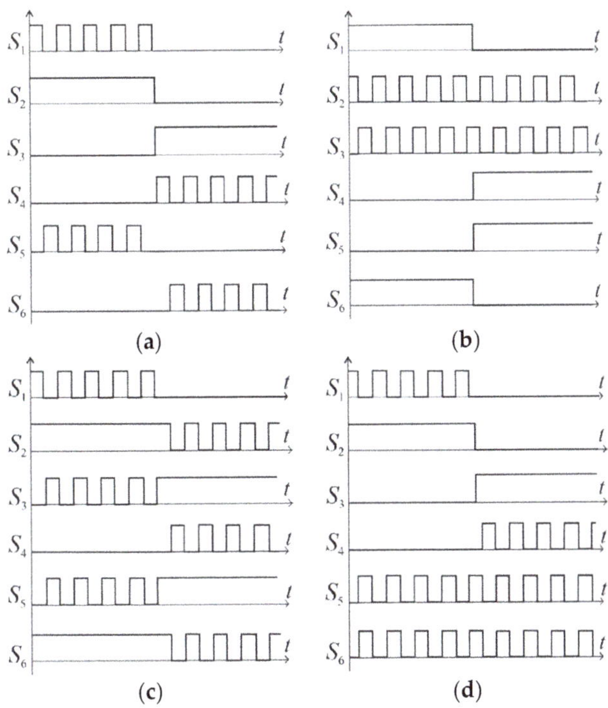

Figure 3. Different modulation strategies for the ANPC converters (**a**) PWM 1, (**b**) PWM 2, (**c**) PWM 3, (**d**) PWM 4.

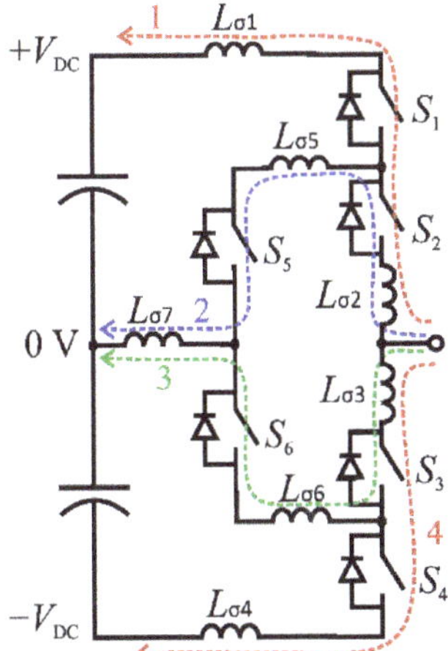

Figure 4. ANPC converter with depicted stray inductances and highlighted conduction paths.

In method PWM1 [34], during zero state transistors, S_2 and S_5 are on, and the current flows through conduction path two (marked in blue in Figure 4). In this control method, during transition P–0 a high value of di/dt in $L_{\sigma 1}$, $L_{\sigma 5}$, $L_{\sigma 7}$ causes voltage spikes on the switching transistors. In technique PWM2 [35], during zero state, the current flows via transistors S_3 and S_6, and in this case during the transition there is a high value of di/dt in stray inductances $L_{\sigma 1}$–$L_{\sigma 3}$, and $L_{\sigma 5}$–$L_{\sigma 7}$. When we compare the transition in PWM2 to the one in PWM1, the equivalent stray inductance is higher, and thus, di/dt is higher as well, which leads to a higher value of voltage spikes. In method PWM3 [23,36], during zero state, transistors S_2, S_3, S_5, and S_6 are on, and current flows through conduction paths two and three; equivalent resistances of both conduction paths are lower, which leads to immensely lower conduction losses. PWM4 [22] is similar to PWM1. In both cases, after the transition from P to 0, the current flows through conduction path two. However, in PWM4 instead of transistor S_6, transistor S_4 is turned on, and thus, even if the modulation pattern differs, the outcome is highly similar.

Generally, the described modulation strategies can be applied using SiC MOSFETs and/or IGBTs in one inverter leg. For hybrid topologies mentioned in Section 1, MOSFETs should be controlled with high frequency, while IGBTs should be switched with fundamental frequency to maximize the system performance. As mentioned before, SiC MOSFETs are more expensive than IGBTs, and thus using different transistors in one module leads to a reduced cost of the converter. However, as SiC MOSFETs become less and less expensive, and the system exhibits better performance with all-SiC configuration. A system with six SiC MOSFETs per leg is thus justified and interesting for further studies, especially including the impact of various modulation techniques.

3. Simulation Study

The system in which the ANPC leg submodule was tested with the different PWM techniques in this paper has been chosen as an open-loop single-phase inverter with a resistive load, as it can mimic the converter's behavior for a power factor near one quite satisfactorily while keeping the circuitry simple. Thus, both simulation and experimental tests were conducted in such a setup, according to system parameters shown in Table 2.

Table 2. System parameters.

Parameter	Description
DC voltage	1500 V
AC voltage	230 V RMS/50 Hz
Rated power	6.67 kVA (1/3 of 20 kVA)
Operating frequency	64 kHz
SiC MOSFETs	NTH4L040N120SC1
Filter inductor	220 µH
Filter capacitor	4.7 µF
DC capacitors	2 × 610 µF

At first, a simulation study in Plecs simulation software was performed to preliminarily showcase the differences in the modulation methods, and establish the power loss split between the converter components. The MOSFETs were modeled based on datasheet values, including the impact of increased junction temperature and different gate resistances. The other crucial source, namely the inductor, has also been included in the power loss analysis, estimating the power loss as a sum of conduction power loss, based on inductor resistance applied in the system based on a real model measurement. Furthermore, component resistances, such as ESR, were added to ensure the converter's loss model is as accurate as possible. However, as Plecs does not simulate the transistor switching processes fully, but rather operates on the basis of a lookup table with power switching loss, the impact of parasitic inductances and ringing could not be observed in the simulation study. Thus, the data from the simulation study were limited to power loss determination and its split between the converter components.

The simulation tests were performed at near-nominal parameters (see Table 2) for all five PWM methods described in Section 2. The impact of the different modulation techniques on the total power loss and its distribution among the converter components based on the simulation study can be observed in Figure 5. In general, for the system parameters, PWM4 is the optimal technique for maximizing efficiency, as the total power loss reached just 114 W, while other PWM methods (PWM1, PWM2, and PWM4) settled close to each other at roughly 137 W.

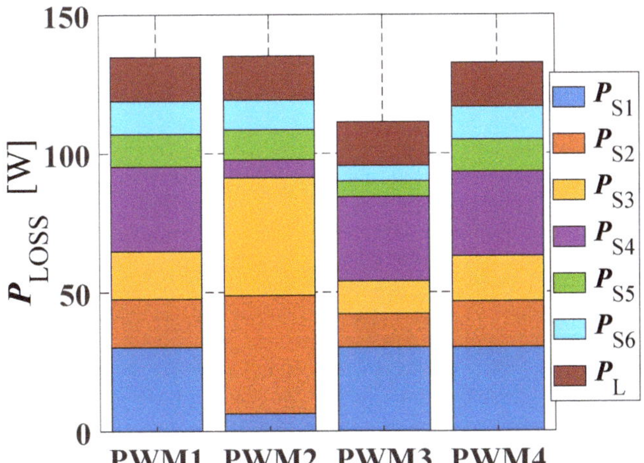

Figure 5. Power loss distribution among the converter components.

When we consider the power loss distribution among the converter components, the situation differs quite notably for the semiconductor power devices, as shown in Table 3, whereas the inductor exhibits nearly identical power loss for all the modulation techniques. At first, PWM1 shows medium total power losses, in which MOSFETs S_1 and S_4 are characterized by the highest value of roughly 30 W each, while pair S_2 and S_3 show 17.3 W, and the last pair S_5 and S_6 just 11.9 W per device. The second modulation method, PWM2, is similar in terms of total power loss. However, it is also characterized by a highly imbalanced distribution—transistors S_2 and S_3 are the sources of over a 70% semiconductor power loss with 42.5 W, while pairs S_1, S_4, and S_5, S_6 emit 6.4 and 10.8 W, respectively. PWM3 exhibits top performance in terms of power loss, with nearly the most imbalanced distribution, as over 60% of the power loss with 30.3 W is dissipated on pair S_1, S_4, while pairs S_2, S_3, and S_5, S_6 are the source of 11.8 and 5.6 W, respectively. However, it is worth noting that the source of this higher imbalance compared with PWM1 and PWM4 is lower power loss for the other MOSFET pairs, and not the increase in the S_1, S_4 pair. Finally, results for method PWM4 are very similar to PWM1 in terms of the loss distribution at 30.3, 16.4, and 11.7 W for transistor pairs S_1, S_4; S_2, S_3; and S_5, S_6, respectively. In terms of the conducting paths and thus power losses, these methods are akin to each other.

Table 3. Power loss distribution among the ANPC submodule transistors.

V_{DC} = 1500 V; P = 6.5 kW; v_{AC} = 230 V rms				
Parameter	**PWM1**	**PWM2**	**PWM3**	**PWM4**
$P_{(S1,S4)}$ [W]	30.3	6.4	30.3	30.3
$P_{(S2,S3)}$ [W]	17.3	42.5	11.8	16.4
$P_{(S5,S6)}$ [W]	11.9	10.8	5.6	11.7

4. Experimental Study

4.1. The ANPC Submodule Prototype

In order to validate the system experimentally, the next step was to design and construct the ANPC leg prototype. Since the system was to operate with a ±750 V DC bus, at least 1200 V rated transistors were needed in the multilevel structure. Based on preliminary calculations from the simulation study and analytical calculations, NTH4L040N120SC1 SiC

MOSFETs were chosen from a group of on the shelf, state-of-the-art power devices as all of the switches, since these are characterized by satisfactory on-state resistance of 40 mΩ and external Kelvin source connection. Therefore, they lead to minimized conduction and switching losses and provide the possibility to switch the transistor in a fast and a robust manner. Furthermore, as mentioned before, it was crucial to minimize the conduction loop lengths in order to lower the effect of parasitic inductances that could lead to excessive ringing, and cause overvoltages and increased power loss, which could result in working outside the safe operating area of the power device and potentially even breakage. This was achieved through the employment of a 4-layer power board structure along with a highly compacted layout of the SiC MOSFETs, as well as additional 82 nF fast bypass capacitors, put between +/0 and 0/− potentials as close to the power devices as possible. Since the plan was to test several PWM methods, none of the conduction paths were favored, and all were of similar length. However, such placement of the transistors leads to a situation where the whole semiconductor power loss has to be dissipated in the near vicinity of the center of the heatsink, thus leading to a less balanced heat distribution; in summary, more capable power loss dissipation measures had to be used. Therefore, as the submodule was to operate with a power of near 6.7 kW in a low-volume system, heatsink Fischer LAM 6 with a highly efficient 48 V fan was employed in the prototype shown in Figure 6.

Figure 6. ANPC submodule—photo of the prototype.

Moreover, the prototype consists of self-made gate drivers based on the UCC21750 chip from Texas Instruments, providing satisfactory switching performance as well as fault protection measures. Finally, the component count of the experimental model of the submodule concludes with two main DC capacitors rated at 800 V and 60 μF. The constructed ANPC leg prototype is shown in Figure 6.

4.2. Experimental Setup

Alongside the ANPC submodule prototype, the experimental setup consisted of an LC line filter, constructed from a 220 μH inductor and a 4.7 μF capacitor, as well as a reconfigurable resistive load. Furthermore, since single-phase systems require high DC capacitance, two additional 550 μF/900 V capacitors were added to support the built-in submodule capacitances. The converter was controlled at 64 kHz operating frequency via a DSP-based circuit, established on TMS320F28379D launchpad. The system was supplied through a 2 kV/5 A DC power supply from Magna Power, and a Yokogawa WT5000 power analyzer was used to measure the efficiency. Finally, the waveforms were obtained using Tektronix MSO56 oscilloscope with isolated voltage probes (Tektronix THDP0100 and P2505A) and a current probe (Tektronix TCP0030A). The scheme and the photo of the experimental setup are shown in Figure 7.

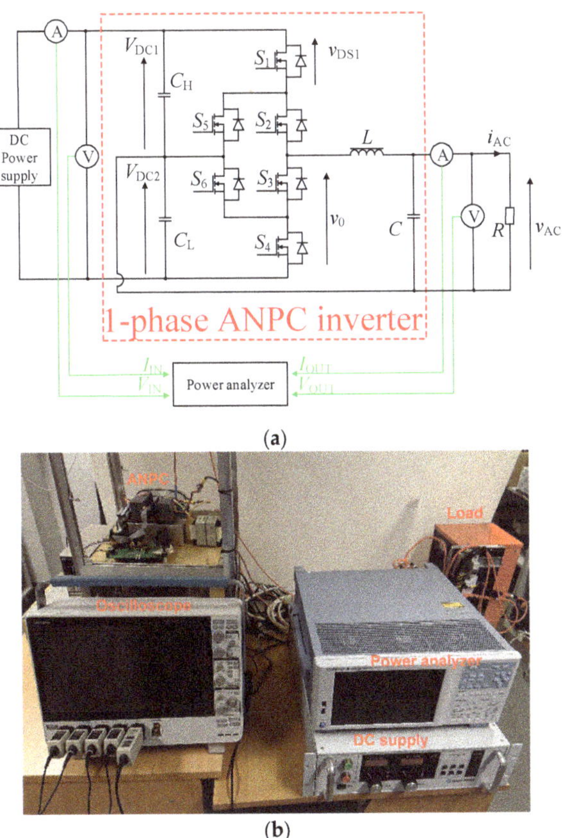

Figure 7. Experimental setup for the ANPC single-phase inverter system with a resistive load—(**a**) scheme, (**b**) photo.

4.3. Results from the Experimental Study

The core focus of the performed tests was to validate the constructed ANPC submodule up to its nominal parameters, and establish the most proficient PWM technique. Since in the whole EV system and the 1500 V DC bipolar bus is required to be connected with the European grid (230 V RMS/50 Hz) at a rated power of 1/3 out of 20 kVA, the modulation index m applied was equal to 0.45, resulting in a voltage gain of roughly 0.153. Figure 8 presents an exemplary oscillogram with line frequency-focused view for PWM3 at 6.7 kW, showcasing ANPC leg output voltage v_0, DC-link voltages V_{DC1} and V_{DC2}, load AC voltage v_{AC}, and current i_{AC}. As the impact of the PWM method is minimal in a 50 Hz context, it is assumed that for other modulation techniques, the waveforms are identical and thus are not shown. As can be seen, there is still some imbalance between the +/0 and 0/− DC voltages, regardless of high 610 μF capacitance; however, its impact is limited regarding the load AC current and voltages as its THD settled below 5% for all the tests, and thus we can omit the mismatches throughout the further result analysis. For the nominal parameters, the AC load current settled at roughly 28 A, whereas the load voltage was established close to 230 V RMS resulting in a power of 6.5 kW. This operating point was further used as a nominal for further experimental comparison between the different PWM methods.

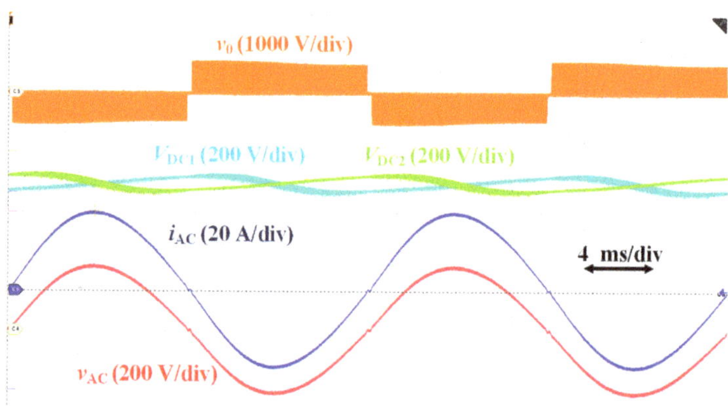

Figure 8. Experimental results from a test at 1500 V DC and 6.5 kW power (m = 0.45, v_{AC} = 230 V) with a line frequency-focused view. From the top: ANPC leg output voltage v_0, DC-link voltages V_{DC1} and V_{DC2}, load AC voltage v_{AC}, and current i_{AC}.

4.3.1. Transistor Overvoltage and Ringing

At first, a study on transistor overvoltage and ringing for different PWM methods was conducted. Since the switching behavior of MOSFETs within the transistor pairs S_1 and S_4, S_2 and S_3, as well as S_5 and S_6 were identical, only the switching voltages for the bottom switches (S_3, S_4, S_6) were measured. Unfortunately, due to the highly compacted design of the submodule, it was impossible to apply current probes and measure the transistor currents. Nevertheless, in terms of the safety of operation for the semiconductor power devices, the drain–source voltage is the crucial factor, while the impact of the current oscillations was indirectly included in the study through efficiency measurements. Furthermore, it is worth noting that transistor overvoltages were also affected by the DC-link voltage imbalances. Thus, peak overshoot voltages could have been even more limited with a higher capacitance, and would not occur if the submodule was used in a different system, e.g., three-phase inverter, where DC-link voltage balancing is assured.

Figures 9 and 10 depict the experimental switching waveforms near the peak line current for the system, operating at nominal values for the modulation technique with the highest voltage oscillations (PWM2). As shown, even though the oscillation is visible, it does not exceed 900 V, which is a safe value for the MOSFETs applied in the system. Furthermore, it is worth noting that due to DC-link voltage imbalance, the waveforms for the positive line current (Figure 9) are different from those obtained for the negative current (Figure 10). This variance between DC-link voltages V_{DC1} and V_{DC2} settled at roughly 60 V, corresponding to roughly 4% of nominal voltage.

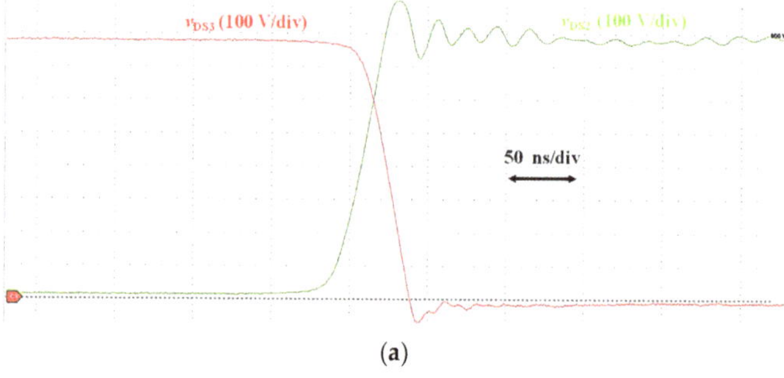

(a)

Figure 9. *Cont.*

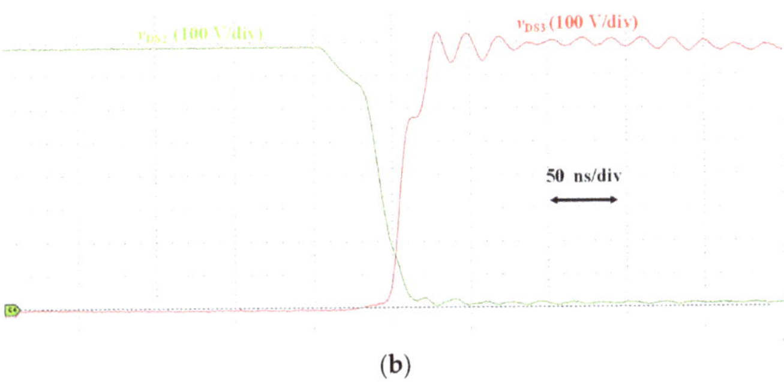

Figure 9. Exemplary experimental waveforms showcasing drain–source transistors for positive load current ($i_{AC} > 0$) for the modulation technique with highest overvoltages (PWM2)—(**a**) turn-on, (**b**) turn-off.

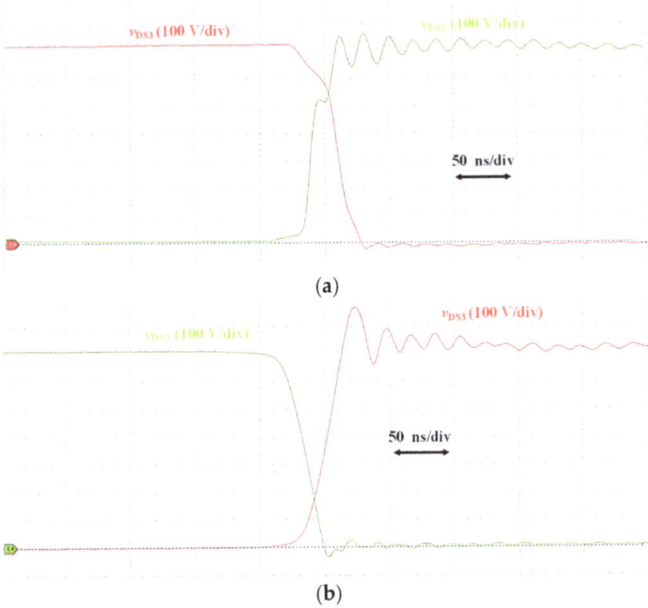

Figure 10. Exemplary experimental waveforms showcasing drain–source transistors for negative load current ($i_{AC} < 0$) for the modulation technique with highest overvoltages (PWM2)—(**a**) turn-on, (**b**) turn-off.

Peak overvoltage values for all the studied modulation techniques are showcased in Table 4. In the previously shown figures, only switching near the peak value of the line current was considered, whereas the data shown in Table 4 consist of the highest value throughout the entire 50 Hz period. Thus, this data are the basis for considerations for all the switches. Based on this data, we can observe that the peak overvoltage value difference between the PWM methods reached maximally 8% of the nominal drain–source voltage of 750 V for PWM2, while the variances between PWM1, PWM3, and PWM4 were as low as 4% of the nominal voltage. The difference between PWM1 and PWM4 is the most visible when transistor voltages are compared. For PWM1, transistors S_1 and S_4 (depending on the line current sign) are not bound to any constant potential, but rather float depending on the current ANPC leg state. This is not an issue for PWM4, as transistors S_5 and S_6 connect the floating potential to the zero voltage, and thus peak transistor voltage overshoots are lower. The significant difference between PWM2 and other techniques is caused by this method's relatively lengthy conduction loop, as mentioned in Section 2. Nevertheless, the variance is still on a minimal level. Therefore, when a similar power rating as in the presented system is considered, it is safe to assume that for a well-optimized system, in terms of conduction path length, the effect of chosen PWM technique on the transistor overvoltage is somewhat limited and should not be as important as other factors, such as power loss and its distribution or filter requirements, or even omitted at all. However, this effect is enlarged when the current is higher. Thus, such an approach should not be applicable in very high power systems.

Table 4. Results for the different modulation patterns at nominal ratings of the ANPC leg.

	V_{DC} = 1500 V; P = 6.5 kW; v_{AC} = 230 V rms			
Parameter	**PWM1**	**PWM2**	**PWM3**	**PWM4**
$v_{DS\_max(S3)}$ [V]	832	957	846	827
$v_{DS\_max(S4)}$ [V]	889	802	868	873
$v_{DS\_max(S6)}$ [V]	857	848	842	843
$P_{LOSS(exp.)}$ [W]	182	177	165	182
$P_{LOSS(sim.)}$ [W]	137	137	114	136

4.3.2. Power Losses

The other crucial factor in which the PWM methods were compared is the system efficiency. Since power loss could not be measured individually on every converter component without impacting the inverter performance, total power loss was measured as the difference between input and output converter power, according to Figure 7.

Figure 11a showcases the total inverter efficiency at nominal system voltages (1500 V DC and 230 V AC), with a constant modulation index at 0.45 and varying load resistance, so that the power could be measured in 20–100% range of its nominal value. As shown, modulation techniques PWM3 and PWM2, similarly as in the simulation study, exhibit the highest efficiency. For the nominal power, and thus with the highest transistor current, PWM4 shows the lowest power losses, implying that the conduction losses are the main source of power loss. This is since this modulation method employs all four middle switches (S_2, S_3, S_5, S_6), and thus, the lowest effective on resistance. PWM2, on the other hand, showcases top performance for lower power, below 60% of the nominal value. In this modulation type, when only the positive half of the line current is considered, transistors S_1 and S_5 do not switch at all as the switching occurs between S_2 and S_3. Therefore, switching loss is limited to this transistor pair, contrary to other PWM techniques where the switching occurs for more power devices. However, when higher power is regarded, the importance of switching loss is diminished, and the efficiency becomes very similar for all the methods except for PWM3, which is characterized by more available conduction paths. Thus, using PWM2 and PWM3 alternately, depending on the load conditions, should be considered to ensure the lowest power losses throughout the whole operating range. Furthermore, developing an algorithm that optimally chooses the modulation technique according to the operating point may be considered. When we study the two other methods more generally, in terms of efficiency, they are pretty similar with the slight advantage for PWM1 for lower power ratings.

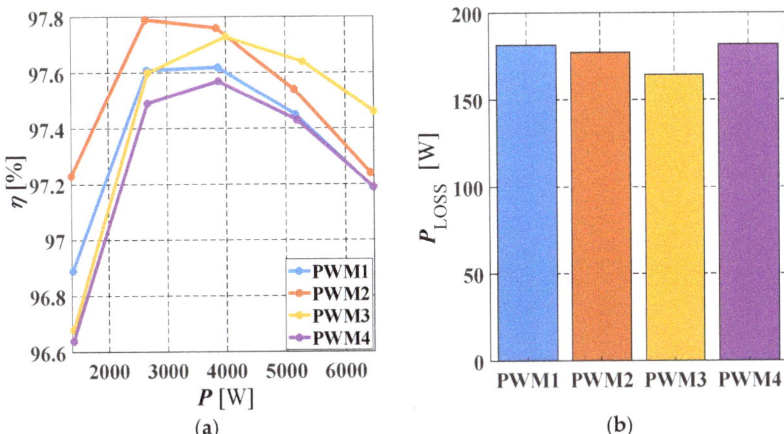

Figure 11. Experimental characteristics showcasing performance of the ANPC leg in function of converter power *P*—(**a**) efficiency at 1500 V DC, 230 V AC, and $m = 0.45$, (**b**) power loss at nominal operating point (1500 V DC, 6.5 kW, $m = 0.45$).

Furthermore, to exhibit the difference in actual loss values rather than efficiency, Figure 11b focuses on the power loss difference for different modulation techniques at the nominal operating point with the full power of 6.5 kW. When we compare the results from the experimental tests with the simulation study (see Table 4) we can observe that the presumptions noted in Section 3 are confirmed via the experiments on the prototype as well, with lowest power losses for PWM3 with 165 W. In contrast, techniques PWM1, PWM2, and PWM4 settled close to each other at 182, 177, and 182 W respectively.

5. Conclusions

This paper presents an MV ANPC submodule with state-of-the-art SiC MOSFETs for an advanced EV charging system. The constructed low-volume prototype of the all-SiC ANPC leg rated at 1500 V DC and 6.67 kW power has been experimentally validated to work with satisfactory switching performance and efficiency above 97.5% for the nominal operating point, which is a substantial value for such a low modulation index and voltage gain (1500 V DC to 230 V AC). Furthermore, as the design process's focus was to minimize the parasitic inductances in the converter, the transistor voltages and ringing were relatively low below 5% of the steady-state value, so that satisfactory switching performance could be achieved. Finally, several PWM techniques have been analyzed, tested, and compared for the specific application shown in this paper, focusing on the impact of all the presented modulation methods strictly for an all-SiC ANPC leg. This is in contrary to other publications in the area, in which the different PWM techniques were applied, but only in various SiC/Si hybrid ANPC leg configurations, usually limited to 1–2 modulations per configuration.

The obtained results show that for an MV all-SiC ANPC inverter submodule rated at 1.5 kV DC and applied in an advanced EV charging system, shown in Figure 1, PWM2 is the best for lower power, while PWM3 is the choice for a higher power (over 60% of nominal value). However, for other power electronics applications, depending on the required voltage levels (and thus the modulation index), as well as for other state-of-the-art SiC MOSFETs and/or Si IGBTs, the outcomes may vary, as so would the ratio between switching and conduction loss, power loss distribution among the components, as well as other factors. Thus, efficiency wise it is not easy to strictly determine which PWM technique is the optimal choice universally, as there are nuances for each application that can affect the power loss quite noticeably. Furthermore, changing the modulation technique during operation, depending on the load parameters should also be considered to achieve the best performance, and thus, the highest efficiency. Nevertheless, based on the theoretical and experimental performed studies for a parasitic inductance optimized system with similar power ratings as in the presented ANPC leg, the impact of conduction loops on transistor overvoltages and ringing between the different PWM methods, even for a quite high voltage of 1.5 kV, is not crucial, and can be nearly fully omitted. Thus, the optimal choice for the modulation technique should be limited to other required parameters, such as efficiency.

Author Contributions: Conceptualization, R.K., M.H. and J.R.; methodology, M.H. and R.K.; simulation study, R.K.; experimental validation, R.K., B.L. and M.H.; original draft preparation, R.K. and M.H.; writing—review and editing, R.K., M.H., R.M. and J.R.; supervision, J.R.; project administration, J.R.; funding acquisition, J.R. All authors have read and agreed to the published version of the manuscript.

Funding: The research leading to these results has received funding from the EEA/Norway Grants 2014–2021.

Institutional Review Board Statement: Not applicable.

Informed Consent Statement: Not applicable.

Conflicts of Interest: The authors declare no conflict of interest.

References

1. Rivera, S.; Kouro, S.; Vazquez, S.; Goetz, S.M.; Lizana, R.; Romero-Cadaval, E. Electric Vehicle Charging Infrastructure: From Grid to Battery. *IEEE Ind. Electron. Mag.* **2021**, *15*, 37–51. [CrossRef]
2. Ghosh, A. Possibilities and Challenges for the Inclusion of the Electric Vehicle (EV) to Reduce the Carbon Footprint in the Transport Sector: A Review. *Energies* **2020**, *13*, 2602. [CrossRef]
3. Ronanki, D.; Kelkar, A.; Williamson, S.S. Extreme Fast Charging Technology—Prospects to Enhance Sustainable Electric Transportation. *Energies* **2019**, *12*, 3721. [CrossRef]
4. Hussain, A.; Bui, V.-H.; Baek, J.-W.; Kim, H.-M. Stationary Energy Storage System for Fast EV Charging Stations: Simultaneous Sizing of Battery and Converter. *Energies* **2019**, *12*, 4516. [CrossRef]
5. Rafi, M.A.H.; Bauman, J. A Comprehensive Review of DC Fast-Charging Stations With Energy Storage: Architectures, Power Converters, and Analysis. *IEEE Trans. Transp. Electrif.* **2021**, *7*, 345–368. [CrossRef]
6. Rivera, S.; Lizana, R.; Kouro, S.; Dragičević, T.; Wu, B. Bipolar DC Power Conversion: State-of-the-Art and Emerging Technologies. *IEEE J. Emerg. Sel. Top. Power Electron.* **2021**, *9*, 1192–1204. [CrossRef]
7. Tahir, Y.; Khan, I.; Rahman, S.; Nadeem, M.F.; Iqbal, A.; Xu, Y.; Rafi, M. A state-of-the-art review on topologies and control techniques of solid-state transformers for electric vehicle extreme fast charging. *IET Power Electron.* **2021**, *14*, 1560–1576. [CrossRef]
8. Poorfakhraei, A.; Narimani, M.; Emadi, A. A Review of Multilevel Inverter Topologies in Electric Vehicles: Current Status and Future Trends. *IEEE Open J. Power Electron.* **2021**, *2*, 155–170. [CrossRef]
9. Bruckner, T.; Bemet, S. Loss balancing in three-level voltage source inverters applying active NPC switches. In Proceedings of the 2001 IEEE 32nd Annual Power Electronics Specialists Conference (IEEE Cat. No.01CH37230), Vancouver, BC, Canada, 17–21 June 2001; Volume 1132, pp. 1135–1140.
10. Ma, L.; Kerekes, T.; Rodriguez, P.; Jin, X.; Teodorescu, R.; Liserre, M. A New PWM Strategy for Grid-Connected Half-Bridge Active NPC Converters With Losses Distribution Balancing Mechanism. *IEEE Trans. Power Electron.* **2015**, *30*, 5331–5340. [CrossRef]
11. Bruckner, T.; Bernet, S.; Guldner, H. The active NPC converter and its loss-balancing control. *IEEE Trans. Ind. Electron.* **2005**, *52*, 855–868. [CrossRef]
12. Schweizer, M.; Kolar, J.W. Design and Implementation of a Highly Efficient Three-Level T-Type Converter for Low-Voltage Applications. *IEEE Trans. Power Electron.* **2013**, *28*, 899–907. [CrossRef]
13. Ronanki, D.; Williamson, S.S. Modular Multilevel Converters for Transportation Electrification: Challenges and Opportunities. *IEEE Trans. Transp. Electrif.* **2018**, *4*, 399–407. [CrossRef]
14. Teston, S.A.; Vilerá, K.V.; Mezaroba, M.; Rech, C. Control System Development for the Three-Ports ANPC Converter. *Energies* **2020**, *13*, 3967. [CrossRef]
15. Lara, J.; Masisi, L.; Hernandez, C.; Arjona, M.A.; Chandra, A. Novel Five-Level ANPC Bidirectional Converter for Power Quality Enhancement during G2V/V2G Operation of Cascaded EV Charger. *Energies* **2021**, *14*, 2650. [CrossRef]
16. Rabkowski, J.; Peftitsis, D.; Nee, H. Silicon carbide power transistors: A new era in power electronics is initiated. *IEEE Ind. Electron. Mag.* **2012**, *6*, 17–26. [CrossRef]
17. Biela, J.; Schweizer, M.; Waffler, S.; Kolar, J.W. Sic versus si—Evaluation of potentials for performance improvement of inverter and dc–dc converter systems by sic power semiconductors. *IEEE Trans. Ind. Electron.* **2011**, *58*, 2872–2882. [CrossRef]

18. Feng, Z.; Zhang, X.; Yu, S.; Zhuang, J. Comparative Study of 2SiC&4Si Hybrid Configuration Schemes in ANPC Inverter. *IEEE Access* **2020**, *8*, 33934–33943. [CrossRef]
19. Zhang, L.; Lou, X.; Li, C.; Wu, F.; Gu, Y.; Chen, G.; Xu, D. Evaluation of Different Si/SiC Hybrid Three-Level Active NPC Inverters for High Power Density. *IEEE Trans. Power Electron.* **2020**, *35*, 8224–8236. [CrossRef]
20. Feng, Z.; Zhang, X.; Wang, J.; Yu, S. A High-Efficiency Three-Level ANPC Inverter Based on Hybrid SiC and Si Devices. *Energies* **2020**, *13*, 1159. [CrossRef]
21. Zhang, L.; Liu, S.; Chen, G.; Yang, X. Evaluation of Hybrid Si/SiC Three-Level Active Neutral-Point-Clamped Inverters. In Proceedings of the 2019 IEEE 28th International Symposium on Industrial Electronics (ISIE), Vancouver, BC, Canada, 12–14 June 2019; pp. 840–845. [CrossRef]
22. Zhang, D.; He, J.; Pan, D. A Megawatt-Scale Medium-Voltage High Efficiency High Power Density "SiC+Si" Hybrid Three-Level ANPC Inverter for Aircraft Hybrid-Electric Propulsion Systems. In Proceedings of the 2018 IEEE Energy Conversion Congress and Exposition (ECCE), Portland, OR, USA, 23–27 September 2018; pp. 806–813.
23. Jiao, Y.; Lee, F.C. New Modulation Scheme for Three-Level Active Neutral-Point-Clamped Converter with Loss and Stress Reduction. *IEEE Trans. Ind. Electron.* **2015**, *62*, 5468–5479. [CrossRef]
24. Floricau, D.; Gateau, G.; Leredde, A.; Teodorescu, R. The efficiency of three-level Active NPC converter for different PWM strategies. In Proceedings of the 2009 13th European Conference on Power Electronics and Applications, Barcelona, Spain, 8–10 September 2009; pp. 1–9.
25. Belkhode, S.; Shukla, A.; Doolla, S. Enhanced Hybrid Active-Neutral-Point-Clamped Converter With Optimized Loss Distribution-Based Modulation Scheme. *IEEE Trans. Power Electron.* **2021**, *36*, 3600–3612. [CrossRef]
26. Mayor, A.; Rizo, M.; Monter, A.R.; Bueno, E.J. Commutation Behavior Analysis of a Dual 3L-ANPC-VSC Phase-Leg PEBB Using 4.5-kV and 1.5-kA HV-IGBT Modules. *IEEE Trans. Power Electron.* **2019**, *34*, 1125–1141. [CrossRef]
27. Jiao, Y.; Lu, S.; Lee, F.C. Switching Performance Optimization of a High Power High Frequency Three-Level Active Neutral Point Clamped Phase Leg. *IEEE Trans. Power Electron.* **2014**, *29*, 3255–3266. [CrossRef]
28. Apeldoorn, O.; Odegard, B.; Steimer, P.; Bernet, S. A 16 MVA ANPC-PEBB with 6 kA IGCTs. In Proceedings of the Fourtieth IAS Annual Meeting. Conference Record of the 2005 Industry Applications Conference, Kowloon, Hong Kong, China, 2–6 October 2005; Volume 812, pp. 818–824.
29. Chen, M.; Pan, D.; Wang, H.; Wang, X.; Blaabjerg, F. Investigation of Switching Oscillations for Silicon Carbide MOSFETs in Three-Level Active Neutral-Point-Clamped Inverters. *IEEE J. Emerg. Sel. Top. Power Electron.* **2021**, *9*, 4839–4853. [CrossRef]
30. Jahns, T.M.; Dai, H. The past, present, and future of power electronics integration technology in motor drives. *CPSS Trans. Power Electron. Appl.* **2017**, *2*, 197–216. [CrossRef]
31. Cittanti, D.; Guacci, M.; Mirić, S.; Bojoi, R.; Kolar, J.W. Comparative Evaluation of 800V DC-Link Three-Phase Two/Three-Level SiC Inverter Concepts for Next-Generation Variable Speed Drives. In Proceedings of the 2020 23rd International Conference on Electrical Machines and Systems (ICEMS), Hamamatsu, Japan, 24–27 November 2020; pp. 1699–1704.
32. Teichmann, R.; Bernet, S. A comparison of three-level converters versus two-level converters for low-voltage drives, traction, and utility applications. *IEEE Trans. Ind. Appl.* **2005**, *41*, 855–865. [CrossRef]
33. Schweizer, M.; Friedli, T.; Kolar, J.W. Comparative Evaluation of Advanced Three-Phase Three-Level Inverter/Converter Topologies Against Two-Level Systems. *IEEE Trans. Ind. Electron.* **2013**, *60*, 5515–5527. [CrossRef]
34. He, J.; Zhang, D.; Pan, D. An Improved PWM Strategy for "SiC+Si" Three-Level Active Neutral Point Clamped Converter in High-Power High-Frequency Applications. In Proceedings of the 2018 IEEE Energy Conversion Congress and Exposition (ECCE), Portland, OR, USA, 23–27 September 2018; pp. 5235–5241.
35. Guan, Q.; Li, C.; Zhang, Y.; Wang, S.; Xu, D.D.; Li, W.; Ma, H. An Extremely High Efficient Three-Level Active Neutral-Point-Clamped Converter Comprising SiC and Si Hybrid Power Stages. *IEEE Trans. Power Electron.* **2018**, *33*, 8341–8352. [CrossRef]
36. Barater, D.; Concari, C.; Buticchi, G.; Gurpinar, E.; De, D.; Castellazzi, A. Performance Evaluation of a Three-Level ANPC Photovoltaic Grid-Connected Inverter With 650-V SiC Devices and Optimized PWM. *IEEE Trans. Ind. Appl.* **2016**, *52*, 2475–2485. [CrossRef]

Article

Interface Converters for Residential Battery Energy Storage Systems: Practices, Difficulties and Prospects

Ilya A. Galkin [1,*], Andrei Blinov [2], Maxim Vorobyov [1], Alexander Bubovich [1], Rodions Saltanovs [1] and Dimosthenis Peftitsis [3]

1 Faculty of Electrical and Environmental Engineering, Riga Technical University, LV1048 Riga, Latvia; maksims.vorobjovs@rtu.lv (M.V.); aleksandrs.bubovics@rtu.lv (A.B.); rodions.saltanovs@rtu.lv (R.S.)
2 Department of Electrical Power Engineering and Mechatronics, Tallinn University of Technology, 19086 Tallinn, Estonia; andrei.blinov@taltech.ee
3 Department of Electrical Power Engineering, Norwegian University of Science and Technology, NO-7491 Trondheim, Norway; dimosthenis.peftitsis@ntnu.no
* Correspondence: gia@eef.rtu.lv

Abstract: Recent trends in building energy systems such as local renewable energy generation have created a distinct demand for energy storage systems to reduce the influence and dependency on the electric power grid. Under the current market conditions, a range of commercially available residential energy storage systems with batteries has been produced. This paper addresses the area of energy storage systems from multiple directions to provide a broader view on the state-of-the-art developments and trends in the field. Present standards and associated limitations of storage implementation are briefly described, followed by the analysis of parameters and features of commercial battery systems for residential applications. Further, the power electronic converters are reviewed in detail, with the focus on existing and perspective non-isolated solutions. The analysis covers well-known standard topologies, including buck-boost and bridge, as well as emerging solutions based on the unfolding inverter and fractional/partial power converters. Finally, trends and future prospects of the residential battery storage technologies are evaluated.

Keywords: residential energy storage; battery energy storage systems; standards; grid interface converters; intellectual property; bidirectional converters; AC-DC power converters; DC-DC power converters; multilevel converters; partial power converters

Citation: Galkin, I.A.; Blinov, A.; Vorobyov, M.; Bubovich, A.; Saltanovs, R.; Peftitsis, D. Interface Converters for Residential Battery Energy Storage Systems: Practices, Difficulties and Prospects. *Energies* **2021**, *14*, 3365. https://doi.org/10.3390/en14123365

Academic Editor: Teuvo Suntio

Received: 31 March 2021
Accepted: 2 June 2021
Published: 8 June 2021

1. Introduction

Consumption of resources as well as their collection and processing are usually uneven. First of the all, it involves energy resources, traditionally, food and various fissile fuels. Nowadays, the necessity to store energy has gained new forms that are applied to the energy resources, specific for the dedicated technology equipment. This, in particular, regards electrical engineering, the rapid development of which during the last two centuries has formed the demand for storages of electrical energy even at the level of residential applications. During recent years, this tendency has become more topical due to several reasons. Firstly, renewable energy sources are in much wider use. In addition, this use is obliged by some administrative regulations like EU directives [1–3]. In spite of the irregular generation profile, the renewable energy sources are being installed even at the households. Secondly, the range and number of various household devices have expanded. There exist plenty of storages dedicated to electrical energy [4]. For example, it is possible to convert electrical energy into chemical (in the form of pure hydrogen) by means of electrolysis and then back—by means of a fuel cell [5]. However, in spite of the most recent achievements in the field of fuel cells [6,7] and development of converter technologies for fuel cells [8], the most functional, reliable and energy efficient equipment for electrical energy is an electrochemical battery energy storage (BES) system.

The constantly increasing number of papers (Figure 1) devoted to battery energy storage systems (BESSs) proves the importance of these energy storage devices in various applications. These papers address all aspects of their use, but particular attention is paid to the interface converters of BESSs. The numerous review papers devoted to this topic [9–12] describe a generalized state of the art in this field. Typically, they evaluate which converter schemes are more energy efficient, with a reduced component count and lower voltage/current stresses. At the same time, the role and peculiarities of the interface converters in the context of the BESS structure are usually not clear-cut and detailed in these reports.

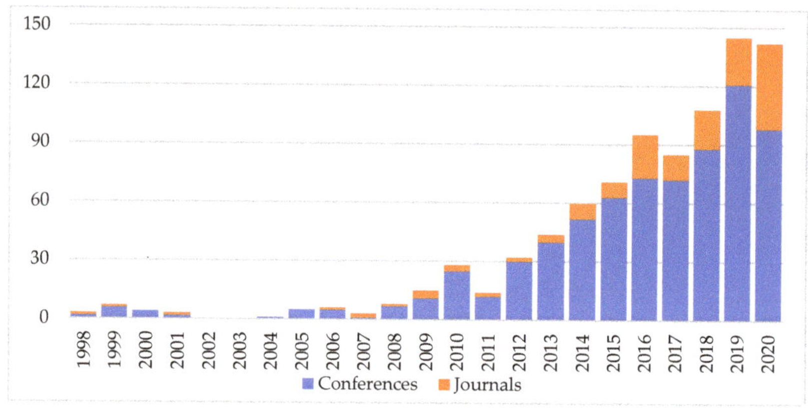

Figure 1. Number of recent IEEE publications about BESS.

BESSs nowadays are also readily commercially available. The analysis of the market of household electrical equipment [13,14] shows that numerous BESSs are already available as a market offering. On the one hand, the variety of their parameters and operation conditions provides wide choices; on the other hand, it makes the choice more complicated for the final users of BESS and complicates the development of the interface converters for different BESSs. In addition, the elaboration and commercialization of BESSs and their interface converters have a strong link to the market of some renewable energy sources and pure electric vehicles, which may not only act as BESSs, but also, after their recycling, provide high voltage (HV) second-life Li-ion batteries for use in BESS [15,16].

The goal of this work is to analyze the majority of interface converters in the context of the corresponding BESSs, their operation conditions (standards, energy tariffs, subsidies and other elements of energy policy), BESS market trends and after this analysis, to formulate prospective development directions of the BESS interface converters. In particular, this regards the converter schemes for HV batteries.

The rest of the paper is organized in five sections. Section 2 reviews the motivating factors of the BESS study: battery technologies, their applications, as well as standards and other regulations that may regard this work. Section 3 briefly analyzes the commercially available BESSs, trying to emphasize their internal structure. Section 4 provides a broad analysis of converter technologies applicable to BESSs. Section 5 discusses the previously analyzed equipment and technologies in the context of BESS development. Finally, the conclusions are given in Section 6.

2. Motivation and Driving Factors for Use of Battery Energy Storage Systems

2.1. Development of Electrochemical Energy Storages

The most intensive development of electrochemical batteries has taken place since the late 20th century and it is still progressing. Due to the constantly growing demand for portable electronics, vehicular technologies and energy systems, the battery technologies of known electrochemistry have been "polished" and new technologies have been introduced to the market. Presently, the most significant commercially available battery technologies are [17,18]: advanced lead-acid (LA), nickel-oxyhydroxide (NiMH), sodium–sulfur (NaS), various kinds of Li-ion batteries, as well as redox flow batteries (RFBs), in particular, vanadium redox batteries (VRBs) [18]. LA technology, the oldest among them, is still the cheapest as well as quite energy efficient (up to 85%). The drawbacks of LA batteries are rather low specific energy (Figure 1) and low number of charge-discharge cycles (lifetime). Historically, the next successive NiMH technology (replacement for NiCd) is characterized by average specific power, specific energy and lifetime, but undergoes significant self-discharge and is of low charge-discharge efficiency (65%). The NaS batteries are of high specific energy, energy efficiency and lifetime (90% and 4000 cycles, respectively [18]), but their operation temperature is high—they require heating, which makes them impractical in many cases. Today, the most quickly developing battery technology is the Li-Ion. Its high specific energy, specific power (Figure 2), lifetime (up to 10k cycles), energy efficiency (up to 95%) achieved at reasonable price makes the technology very suitable for use in portable electronics, all-electric vehicles, household energy systems, and, even, in energy distribution grids [19]. However, the specific parameters of Li-Ion batteries depend on relevant chemistry and all advantages are typically not concentrated in one device. Finally, RFBs, in particular VRBs, are the batteries that utilize reduction–oxidation reaction between two liquids, which occurs through a membrane. The liquids are pumped to the membrane that makes RFBs similar to fuel cells, where the liquids are chemically restorable. The main advantage of these batteries is their potentially infinite lifetime. Lastly, it must be mentioned that modern batteries are not just a series connection of galvanic cells. They often include electronics for balancing, management and protection as well as chargers in some cases. Therefore, these batteries can be considered as complex complete energy units for immediate use [20–22].

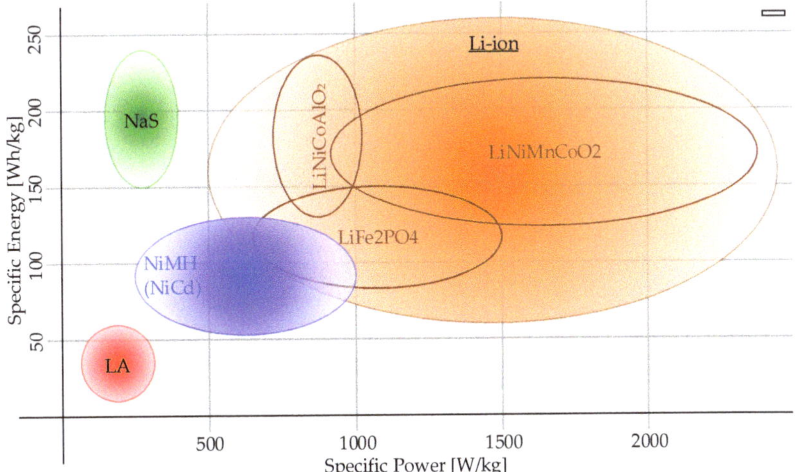

Figure 2. Specific energy and specific power of the commercially available batteries (based on data from [18]).

2.2. Extensive Use of Battery Energy Storages in Transport

One of the recent global societal and legislative tendencies on the national and international levels is the request to reduce the consumption of fossil fuels and to increase the efficiency of energy consumption [3,23]. Among other areas, this involves vehicular technologies as well. Regarding ground vehicles, this initiative means wider use of plug-in electric vehicles (PEVs) or all-electric vehicles (AEVs) and hybrid ones, equally in the public and private sectors. In [24,25], the availability of cost-effective batteries of several hundred volts for main electrochemical energy storage of PEV is reported. More recent papers [26–29] consider these PEV storage systems valuable enough to be a part of the energy supply grid. Further development of the BESs makes their use possible in larger ground vehicles—first of all, in the public transport [30,31].

Better BESs are also required for water vehicles, first of all, for smaller auxiliary vehicles—boats, yachts, water buses, etc. For example, in [32], the electrification of the water buses in Venice is considered as a successful example of BES use in water transport. At the same time, with regard to bigger ships and vessels, the role of BESs differs with time. While earlier configurations of marine energy systems utilize high voltage batteries for stabilizing the traditional on-board AC grid and power smoothing [33,34], modern systems also take into account the possibility of all-electric propulsion of the ship [34,35].

Finally, the most advanced BESs are applicable in aircraft. The traditional electrical supply of an aircraft combines an AC and DC grid. Better performance of the applied batteries leads to a better quality of the 28 V DC grid [36,37]. At the same time, top BES technologies allow production of extremely light batteries that enable all-electric aircraft [38].

In conclusion, the extensive use of batteries in transport, in particular, the growing number of light PEVs, high capacity of their batteries and huge capacity of these batteries in total, as well as their wide distribution, make these BESs a substantial grid resource for storing energy. These BESs and their interface converters are typically high-voltage devices, but the corresponding solutions of the interface converters can be adopted for residential use.

2.3. Recent Challenges in the Field of Power and Energy Supply

The request to reduce fossil fuel consumption [3,23] regards also power distribution and supply networks. For the power and energy supply systems, this means that the burning of fossil fuels must be substituted with renewable energy generation. In turn, the main properties of renewable energy generation are:

(1) Uneven generation profile—regardless of the kind, the renewable energy sources typically do not provide constant power. In particular, the generation of PV panels depends on solar irradiation and varies with the daytime, cloudiness, season, location of PV and solar activity. The generation of wind turbines depends on the wind strength, which is unique for its location, season and occasional weather fluctuation. The generation of hydro and waves turbines depends on the amount of water that is a long-term function of seasonal and global weather changes.
(2) Variety of power ratings and types of energy sources exist even within the same group. For example, the power of PV depends on the local properties and financial abilities of a particular household.
(3) Variety of allocation of the renewable energy sources—depending on the particular economic conditions and policy of energy operator, these sources may be allocated differently.

Altogether, this makes renewable energy generation less stable and reliable. This, as well as several other problems [9,39–43], can be solved with the help of Battery Energy Storage Systems (BESSs).Figure 3 shows the use of BESSs in energy applications.

When considering a BESS in a small household with different loads and renewable energy sources, it is very important to smoothen renewable energy generation—providing storage for excessive renewable or cheap grid energy [44–46]. The BESS is also capable of performing the function of an uninterruptible power supply. This is the main function in the case of islanded residential grids [47–49]. At very uneven loads, the BESS may also smoothen the real-time loading of supply equipment—transformers and lines.

In the distribution grids, the functions of the dedicated BESS are similar but more specified. Price compensation now can be considered as a complete function of energy trading, smoothening of power generation regards not only renewables, but smoothening of consumed power at this level saves the capacity of distribution equipment. Additionally, BESS in distribution grids may perform grid service functions: grid black restart as well as voltage and frequency regulation [39,50]. The choice of BESS parameters is a subject of multiple factors [51]: standards, power losses, voltage of majority of available PEVs, compatibility with pure resistive loads.

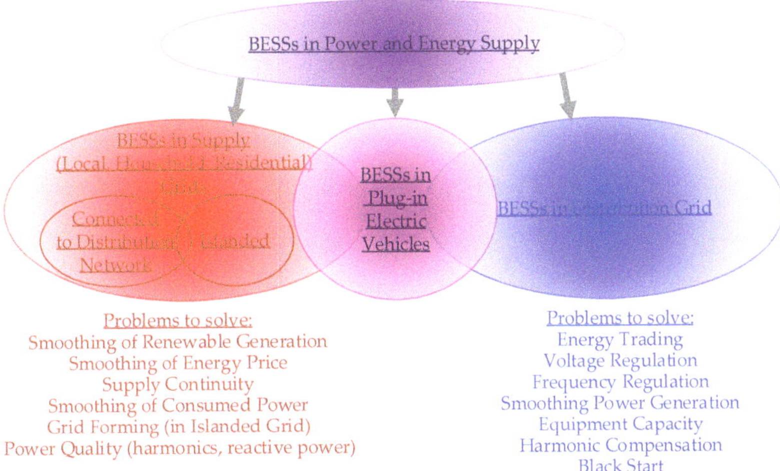

Figure 3. Battery Energy Storage Systems (BESSs) in power and energy supply at a glance.

2.4. Standards and Other Regulations Applicable to Battery Energy Storage Systems

The standards directly related to the electrical energy storage systems of households are still under development. In Europe, this is being done by the IEC 120 committee group [52]. They have developed a roadmap for developing standards, which is planned to be completed by the end of 2023. Until that date, European manufacturers have to use general standards for the production of power converters, in particular, power interfaces for alternative energy sources and uninterruptible power supply (Table 1).

Table 1. Summary of Standards and Regulations applicable to BESS.

Reference	Application Area of the Standard
[53]	USA, Converter housing and selection of components
[54]	IEC, Classification of BESS locations in households
[55]	IEC, Voltage inverters for high voltage DC networks
[56]	IEC, Controlling of converters in microgrids
[57]	IEC, Connection of PV to the grid and requirements for electromagnetic compatibility parameters
[58]	IEC, Bidirectional low voltage (up to 1000 V AC and 1500 V DC) converters connected to the grid and description of the terms used in these networks
[59,60]	IEC, Test methods and acceptable parameters for low voltage uninterruptible power supplies
[61]	IEC, Disposal of converters of uninterruptible power supplies
[62]	USA, Safety regulations within data centers and telecom central offices

In the USA, a universal standard has been developed that describes the operation of electrical energy converters in distributed networks. With regard to BESS, the manufacturers also have to apply general standards for converters. This includes standards for interface converters of energy storage. In addition, in the USA, the parameters of batteries are defined and standardized and based on the standards of telecommunication equipment (Table 1).

3. Commercially Available Residential Storage Systems

In this section, the BESs available on the market are analyzed taking into account the parameters available from the product datasheets or application manuals. Despite the market for such devices still being dynamic, some common properties and features can already be distinguished as common practice in the field.

3.1. Typical Example of Battery Energy Storage Systems Dedicated to Household Applications

The Tesla Powerwall 1 (3.3 kW/6.4 kWh) was one of the first attempts to include BESS into a household energy system and has been available on the market since 2015. It operates with a DC-bus and, in general, has to be installed in conjunction with a grid inverter, which is sold separately.

This precluded its use as a completely independent BESS, reduced market prospects and shortly led to its replacement by the Tesla Powerwall 2 (5 kW/13.2 kWh) [63]. In contrast to the previous model, the Powerwall 2 (Figure 4a) includes an AC inverter and can be connected directly to the AC grid. This enables its use as a residential BESSs, regardless of the renewable generation source (solar panels or a wind generator). Therefore, the functional features of Powerwall 2 have expanded significantly, including the possibility of stand-alone operation without grid connection (islanded mode). For normal operation, it requires an additional commutation unit called "energy gateway" and its full cycle efficiency is 90%.

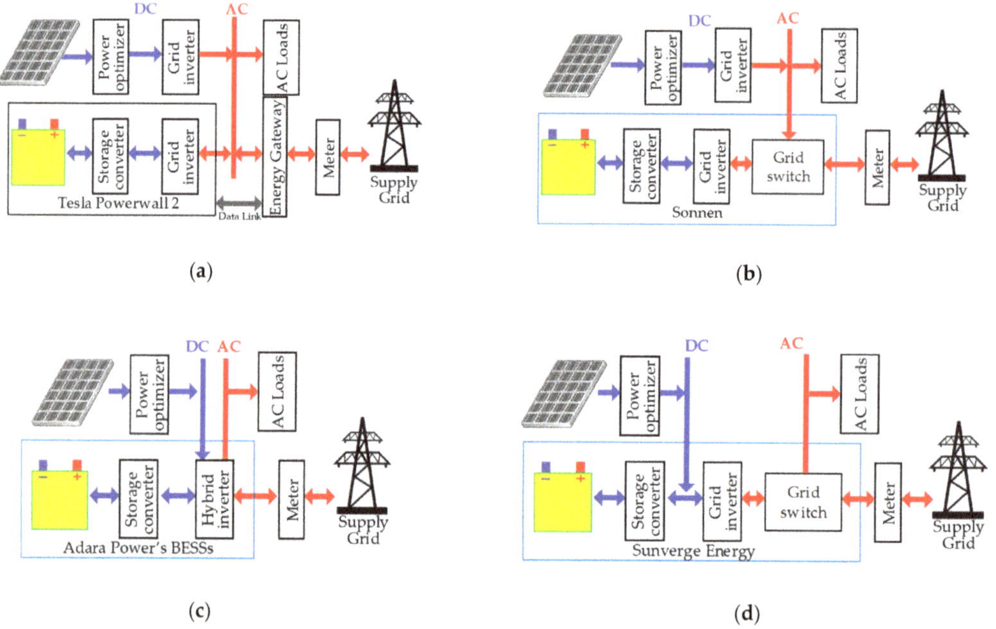

Figure 4. Typical examples of system configurations of different BESs: (**a**) Tesla Powerwall 2, (**b**) SonnenBatterie, (**c**) Adara Power-Residential and (**d**) Sunverge energy.

Sonnen is another early market player that began offering its residential BESSs in December 2015 [64]. These BESSs are designed for households with solar and wind power generators providing energy storage and backup power. They are available in two versions, with a built-in inverter for PVs (hybrid output) and without it (eco output). In Germany, the company launched a coordination network that brings together power producers and storage owners. This service allows the participants resided in the same network to exchange electricity with each other, exporting surplus to the grid. Currently, this service has over 10,000 users. With $LiFePO_4$ batteries in its system, the manufacturer claims an output power of 2.5–3.3 kW in the "eco output" version and 5.5 kW in the "hybrid output" version. The energy capacity of the base model is 5 kWh with the ability to increase it up to 15 kWh in 2.5 kWh steps. The manufacturer promises a 98% maximum efficiency of the batteries and a 96% efficiency of the converter, which gives a total cycle efficiency of around 88.5%. The internal topology of SonnenBatterie and SonnenFlat is not disclosed, but the structural diagram of their operational environment (Figure 4b) shows that it is connected to the main supply grid as well as to the secondary grid formed by the solar panel inverter through an automatic transfer switch (ATS). This enables a SonnenBatterie to operate in an uninterruptible power supply mode.

Enphase Energy is another company that entered the residential storage market with its "AC Battery" in 2015. It is a very compact (0.27 kW/1.2 kWh) modular system that can be used in conjunction with micro-inverters and the "Envoy-S gateway" [65]. Later, the company's storage portfolio was extended with the Encharge 3 (1.28 kW/3.3 kWh) [66] and Encharge 10 [67], which is composed of three of the former units. According to the datasheet information, the cycle efficiency of a newer Enphase product is 89% at half power. Backup power from the battery can be provided using an additional microgrid interconnection device.

Other notable market players are Victron Energy with a range of products like Easy Solar and MultiPlus [68]; Adara Power's Residential [69,70] coupled with an inverter from Schneider Electric (Figure 4c) [71] and Sunverge Energy (Figure 4d) [72]. Moreover, one of the key market players is the battery manufacturer LG Chem [73], who is offering its low- and high-voltage battery modules for integration with SMA, Fronius, SolarEdge, and Huawei inverters/chargers.

3.2. Summary of Parameters and Features of Commercial Residential BESs

Due to the market dynamics, with both large and small companies are entering and leaving the market continuously, so it is hard to determine a global leader in the area. Moreover, some of the products currently have a limited proposal or are available only in certain regions. The typical price for typical residential BESs is currently in the range of 1–2 kEUR/kWh (Table 2). Technical information on these products is mostly limited—only general specifications are typically available. Still, certain common properties of residential BESSs can already be distinguished. In the majority of cases, the utilized energy storage is a low-voltage (50 V) Li-ion battery, which is associated with relatively high currents. Although the particular topology configurations used in these systems are not revealed by the manufacturers, such voltage level would in general require a rather complex interface converter featuring a transformer for the required voltage step-up. Using RESU10 and RESU10H from LG as a reference, one of the reasons for using a battery with such voltage level is its 14% reduced price, as compared with the higher voltage battery of the same energy capacity. This results in round-trip efficiencies of most residential BESS being around 90%, which seems to be a current technological limit for such configurations.

The current market of BESSs shows a clear trend of their transformation from the auxiliary BESSs, complementing a solar or wind farm with a smoothing energy storage (AC and DC coupling), towards a complete energy system with BES that does not depend on the availability of alternative energy (only AC coupling). While the BESSs of the first type have a DC output and often need a separate grid converter, newer BESSs of the second type are intended for AC operation due to the intrinsic AC interface. From the point of view of their features and functions, the earlier BESSs were focused on local power supply and equalization or shift of peak consumption, but the newer systems have a range of advanced functions, for example, integration on energy system level, i.e., possibility of combining several household grids with BESS into a distributed power plant.

Table 2. Summary of BESS for general use and use with renewable energy sources available on the market.

BESS Manufacturer/Model	Maximal Energy Capacity [kWh]	Charge/Discharge BES Power [kW]	Battery Voltage [V]	Coupling	Reference
Tesla PowerWall	13.5	5	50	AC	[63]
Sonnen Batterie Eco	15	3.3	48	AC	[64]
Adara Power (Residential)	20	12	50	AC/DC	[70]
Sunverge	Modular up to 19.4	6	48	AC/DC	[72]
Solax X-ESS G4 or Hybrid X1/X3 + Triple Power (BES)	Stackable up to 23 (4 modules)	4	300	AC/DC	[74]
SolarEdge + RESU10H	9.8	5	400	AC/DC	[75]
PowerVault 3	20	3.3/5.5	52	AC	[76]
Puredrive Storage II AC 5 kWh	5/10	3	50	AC	[77]
Duracell Energy Bank	3.3	3.3	52	AC	[78]
Enphase Encharge 3	3.5	1.3	67	AC	[79,80]
Enphase Encharge 10	10.5	3.8			
Nissan/Eaton xStorage	4.2 … 10	3.6 … 6	90	AC/DC	[81]
Samsung SDI All in One	3.6	4.6	60	AC/DC	[82]
Varta Pulse/Pulse Neo 3	3.3	1.6/1.4	50	AC	[83]
Varta Pulse/Pulse Neo 6	6.5	2.5/2.3			
Sunny Boy Storage	External battery	3.7/5/6	360	AC	[84]
Victron Energy EasySolar	External battery	0.9/1.7/3.5	12.8–51.2	DC	[85]

3.3. *Isolated Converters of Commercially Available Residential BESSs*

As it was shown in Sections 3.1 and 3.2, most of the commercially available BESSs utilize a low voltage battery (see Table 2 for details). The use of such a low-voltage battery while maintaining, at the same time, good control performance, requires that the entire BES interface converter or part of it be a low voltage circuit that, in turn, typically means the use of an isolation transformer. The use of the transformer also allows satisfying the potential safety requirements (see Section 2.4 for details). The transformer may be a network transformer operating at the frequency of the supply grid or a high-frequency transformer. Both solutions have benefits and disadvantages briefly considered below.

3.3.1. Converters with Grid-Frequency Isolating Transformers at AC Side

In general, adding of a transformer at the grid side moves the entire semiconductor circuitry into a low voltage operation, but its topology may be almost of any type, as presented below in Section 4. Therefore, there are two large groups of converters with a network transformer: single stage converters (Figure 5a) and converters with two conversion stages (Figure 5b). The BESS may also be equipped with a transformer at the request of the operator and/or legal regulations, in order to meet the operational requirements. This, however, regards more to BESSs for distribution grids, in particular, ABB with the ESSPro product line [86] and NIDEC with the Silcolstart product line [87]. The transformer installed at the AC side makes the operation of the converter possible at lower voltages, but makes the BESS heavier and bulky.

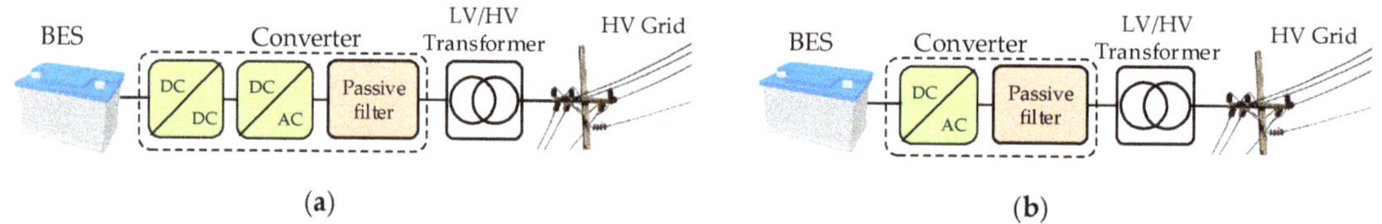

Figure 5. BESS power electronics converters with a transformer: (**a**) single-stage, (**b**) two-stage.

3.3.2. Converters with High-Frequency Isolating Transformers

A high-frequency transformer may be allocated in the DC-link (Figure 6a). The most versatile and straightforward kind of the implementation of this approach is the use the circuitry known as dual active bridge (DAB, Figure 6b). Classical DAB [88] is a hard-switching topology that, compared with non-isolated interface converters, considered below in Section 4, is less reliable and energy efficient due to the extra components as well as due to its hard-switching nature. However, if combined with a soft-switching technique, for example, applying a resonant network, it may operate with better efficiency [89,90].

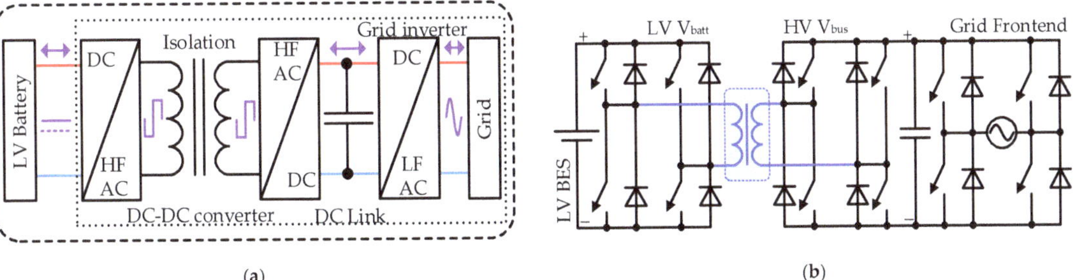

(a) (b)

Figure 6. High-frequency transformers in the DC-link of the interface converter of LV BES: (**a**) functional diagram, (**b**) transformer in conjunction with classical DAB.

In a more advanced approach (Figure 7a), the high-frequency transformer is located at the edge of the DC-link and the AC grid [91]. This requires that the AC part of the topology contains bidirectional switches so that it can operate at both polarities of the grid voltage. The performance of this topology can be improved with the help of resonant networks (red elements in Figure 7b) and advanced modulation methods [92]. Similar topologies and their properties are well described in [93].

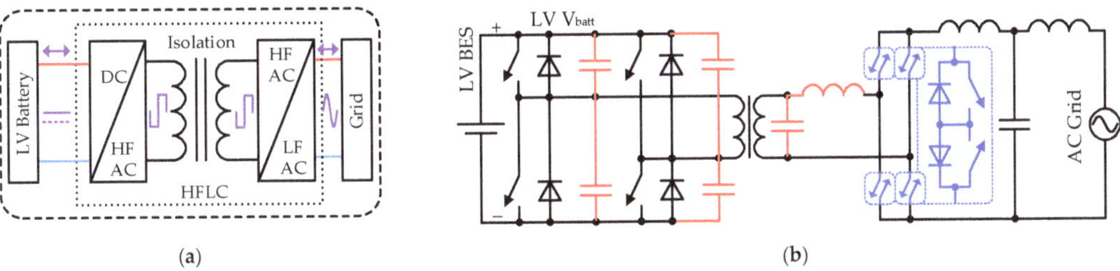

(a) (b)

Figure 7. Full-bridge DC-AC converter with a high-frequency transformer: (**a**) functional diagram, (**b**) converter derived from DAB.

The abovementioned DC-DC and DC-AC converters contain an energy-bypassing transformer. Alternatively, the high frequency link may contain also a storage transformer (split coil). In the most explicit form, this storage transformer is seen in a flyback converter. This converter, however, is a DC-DC circuit and its use, therefore, is directly possible only in the DC-link [94] similar to DAB (Figure 6b). At the same time, adopting of the principle to AC networks is possible. For example, the converter presented in [95] contains two flyback converters dedicated to positive and negative half-waves. The inputs of the converters are connected in parallel to a low voltage battery, but their outputs—in series to the grid (Figure 8). The interface converters with a high-frequency storage transformer have the same drawbacks as original flyback converters: rather low power and highly pulsating current on both sides (including battery side) that requires sufficient filtering.

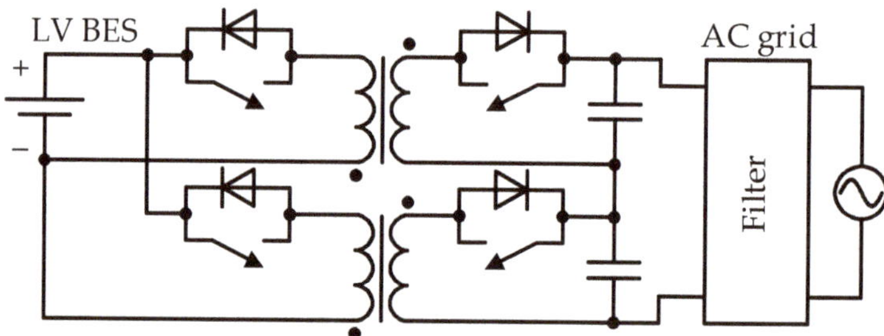

Figure 8. Inverter with a high-frequency transformer derived from a flyback converter.

4. Topologies of Non-Isolated Interface Converters for High-Voltage Battery Energy Storage Systems

One of the ways to overcome some limitations of the existing residential BESS is to utilize a battery with higher voltage (~200–500 V) and enable the use of a simpler and more efficient interface converter. In fact, some companies, like SolaX, SMA and SunnyBoy, are already on this path. Due to massive electrification of transportation where higher voltage batteries are used to reduce charging current and time, the cost for higher voltage batteries should decline further and make the use of high voltage (HV) batteries more feasible for residential BESSs.

This section is devoted to the analysis of existing and perspective non-isolated power electronic interfaces that can be applicable to the residential HV BESs. The main goal is to highlight the benefits and limitations of various configurations and assess their feasibility and performance. In addition to the standard single-, two-stage and multilevel topologies, emerging configurations like impedance-source, partial and fractional power converters are analyzed.

4.1. Functions and Structure of Interface Converters for BES

According to the analysis of commercially available residential BESSs, two main configurations can be distinguished: DC- and AC-coupled. The first group is generally represented by the power electronic systems that are often referred to as "hybrid inverters" (Figure 9a). They allow integration of both PV and battery into a single multiport unit. Such solutions are well-suited for new installations, but the choice of suitable storage configurations could be limited. On the other hand, the AC-coupled storages are often stand-alone systems that are directly connected to the residential AC grid (Figure 9b). In general, such solutions are more flexible, as they can be integrated into any existing installation. However, for such systems, charging of a battery from a PV typically involves more energy conversion stages, with a negative impact on efficiency.

The interface converter of a BES needs to perform two main functions, along with a range of auxiliary application-based functions. The main functions of the BES are sinusoidal shaping of the AC grid current and forming the DC current of the BES in both directions of power flow. The abovementioned functions can be implemented in a single stage bidirectional DC-AC inverter/rectifier; however, such solutions are typically overall less efficient due to battery voltage variation as compared to two-stage systems [96]. Therefore, the BES interface is usually comprised of a bidirectional DC-DC stage that is interfaced with a battery, followed by the DC-AC inverter/rectifier. The state-of-the-art and other potential configurations of power electronic interfaces for HV BES are analyzed in the following sections.

4.2. Single Stage DC-AC Bidirectional Inverters/Rectifiers

This section presents the state-of-the-art and emerging single-stage grid-tie inverter/rectifier topologies. The main goal of these systems is to convert a DC voltage into the sinusoidal AC waveform and vice versa. Most commercial systems require the

DC voltage to be relatively stable, with their value higher than the amplitude of the grid voltage, while some of the emerging topologies potentially offer enhanced flexibility.

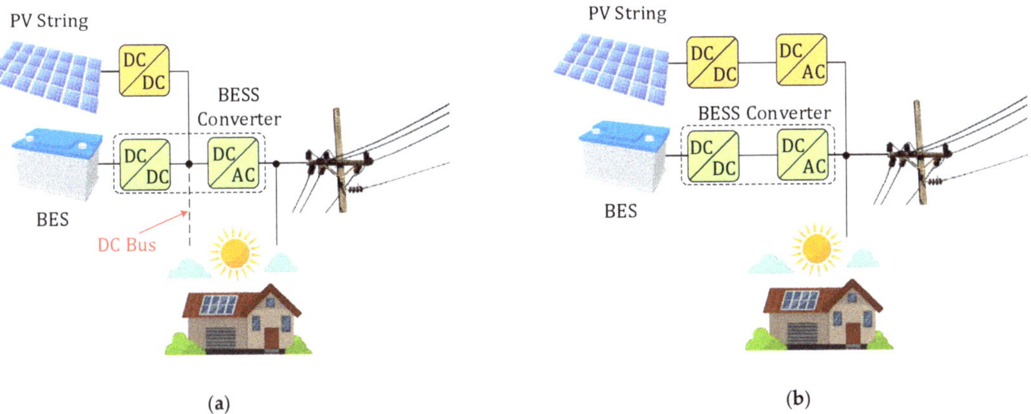

Figure 9. Coupling of units in residential energy systems: (**a**) DC coupling, (**b**) AC coupling.

4.2.1. Bridge Converters

A high voltage battery can be attached to the grid through a single stage bidirectional (four-quadrant) interface converter. The most frequently mentioned converter is a transistor bridge. Such bridge itself is a parallel connection of two (three) transistor legs with two transistors (and anti-parallel diodes) in each (Figure 10). The converter includes also an inductance coil between its AC terminals and the grid implementing an AC current source (Figure 10a). A diagonal couple of transistors and the couple of diodes located in the opposite diagonal form a chopper capable of converting the grid voltage at its particular polarity. In Figure 10a, red elements represent the chopper for the positive half-wave, but blue elements—for the negative. The chopper is bidirectional and can be considered as a buck converter supplied from the DC bus or as a boost converter supplied from the AC grid [97]. One transistor leg can be substituted by a series connected capacitors (capacitor leg), thus forming a transistor-capacitor bridge (Figure 10b), more frequently named "half-bridge" [97]. In this topology, it is also possible to identify two choppers for both half-waves of AC voltage. Finally, it is possible to apply this approach of schematic synthesis to three phase systems (Figure 10c,d). This forms the three-phase transistor bridge and the three-phase transistor bridge with a capacitor leg coupled to the grid via inductor-based AC current source [98].

4.2.2. Topologies without Explicit Bridge

The intrinsic choppers shown in Figure 10 can be deployed without forming an explicit bridge. This scheme is defined as a dual-buck grid converter known since 1997 [99,100]. With this approach, the elements of the "positive" and "negative" choppers are different, which enables them to be further optimized.

Figure 11 shows how the elements of implicit choppers are extracted (red elements—for positive and blue elements—for negative). With this approach, the inductance coil is not shared between "positive" and "negative" branches (Figure 11a). These coils can be magnetically coupled (Figure 11b), providing their lower weight/volume and therefore, lower weight/volume of the converter itself [101], without the reduction of the performance and reliability of the converter. Alternatively, both branches can be combined through a couple of series connected diodes (Figure 11c) [102], keeping the same advantages.

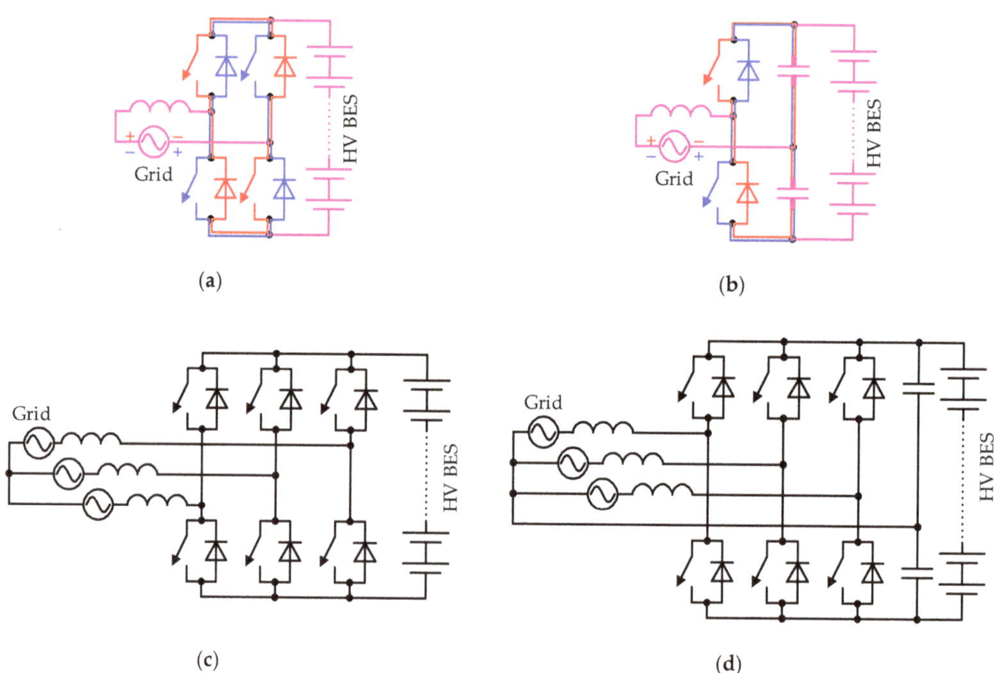

Figure 10. Configurations of single-stage bridge rectifiers-inverters for BESs: (**a**) AC current sourced transistor bridge, (**b**) AC current sourced transistor/capacitor bridge (half-bridge), (**c**,**d**) three-phase schemes.

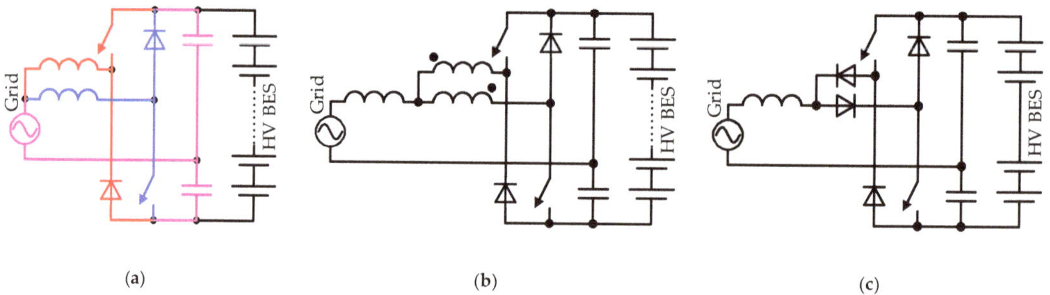

Figure 11. Dual-buck grid converter: (**a**) derived from half-bridge, (**b**) magnetically coupled branches, (**c**) diode coupled.

Extracting of the intrinsic voltage converters at both grid terminals of a full-bridge converter forms another kind of the dual-buck grid converter (Figure 12a) [102]. Another version derived from the full-bridge topology can be synthesized by means of direct combining of two DC sourced buck converters—attached to each terminal of the grid [76]. In this converter, the switches located at the grid side are continuously conducting at the corresponding grid voltage polarity that reduces switching losses. Additionally, such converter may be "tied to positive voltage node" (as shown in Figure 12b) or "ground tied". Finally, adding two diodes at the grid side (Figure 12c) allows operating in "ground tied" and "positive node tied" modes [103,104], making the operation of the switches more symmetrical. The converter shown in Figure 12b can be equipped with magnetically coupled inductance coils or coupling diodes, as shown in Figure 11.

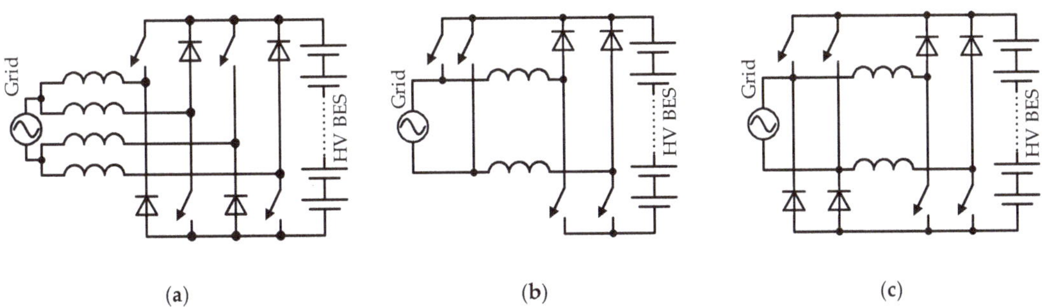

Figure 12. Dual-buck grid converters: (**a**) derived from full-bridge, (**b**,**c**) synthesized of two combined buck converters.

Due to a low number of elements, the considered single-stage converters operate with best efficiency at their particular operation point. However, when considered in conjunction with the attached battery, their efficiency is not outstanding and drops significantly at other operation points due to the higher losses in the converter and the battery [86]. In addition, these converters operate as an AC grid supplied boost or a BES supplied buck converter that requires minimal battery voltage to be higher than the amplitude of the grid voltage.

4.2.3. Multilevel Converters

Multilevel converters (MLC) can be considered as a specific kind of the single-stage converters, processing energy in separate cells of a BESS battery. In contrast to the above-considered topologies that always deal with the same DC voltage or with the entire battery, the multilevel converters form their output of DC voltage that may have several levels obtained directly from the battery. The advantages of multilevel converters are lower harmonic distortion, switching losses and electromagnetic interference [105]. There are three main topologies of multilevel converters: cascaded H-bridge converters (also known as multilevel converters with independent sources), neutral point clamped multilevel converters (also known as diode clamped multilevel converters), and multilevel converters with flying capacitors.

Cascaded H-Bridge Converter Structures

In the case of cascaded H-bridge multilevel converters, each phase contains several series-connected modules (Figure 13a) composed of dedicated cells and an inverter, together forming an independent source. Within a BESS [106], these sources can be charged and discharged more evenly due to the independent nature of their involvement in the current path and potentially free exchange of the sources [107].

There exist various types of power converters and energy storage building blocks. The most common converter is a single-phase transistor bridge (H-bridges) shown in Figure 14a, which generates AC voltage on its output, thus controlling the charge or discharge process of the connected battery cells. Another typical configuration given in Figure 14b includes an AC generating H-bridge in conjunction with a synchronous buck converter that compensates voltage changes in the cell(s).

Lastly, a successful commercial implementation of BESS with a multilevel converter was offered by SolarEdge [108]. It is based on a multilevel DC converter with multiple DC modules connected in series (Figure 13b). Allocation of the multilevel structure in the DC bus enables significant simplification of the cell converters (Figure 14c). The DC/DC converters can operate in the following modes: balancing circuit, charger and battery discharger. In turn, if the DC bus is formed by an MLC, the grid frontend can be a simple commutation matrix or an efficient pulse mode inverter or a short-circuit proof converter with an impedance network. A similar topology developed by ABB for distribution networks [109] includes an array of complex cells containing two transistors and a battery with switches and capacitors. A cell may work as a boost or buck converter and is capable of shunting the cell if needed.

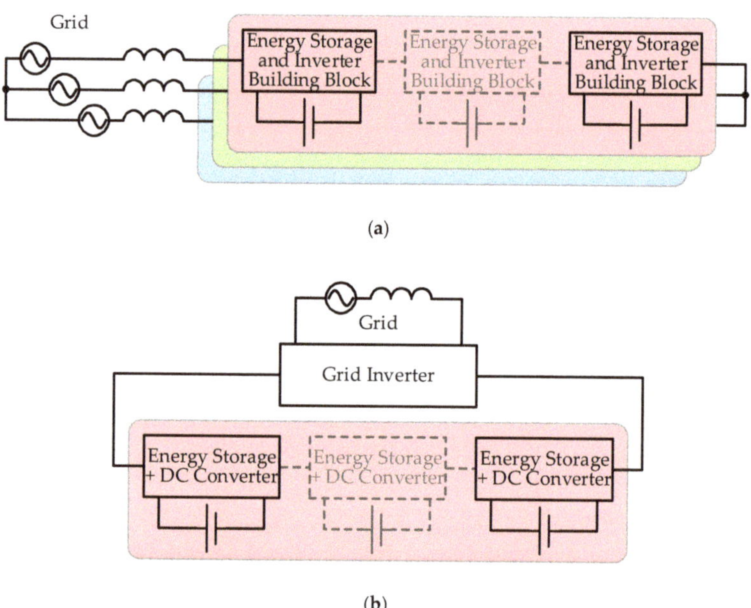

Figure 13. Functional diagrams of cascaded H-Bridge multilevel converters: (**a**) traditional configuration of AC MLC, (**b**) configuration DC MLC with unipolar cell converters and grid frontend.

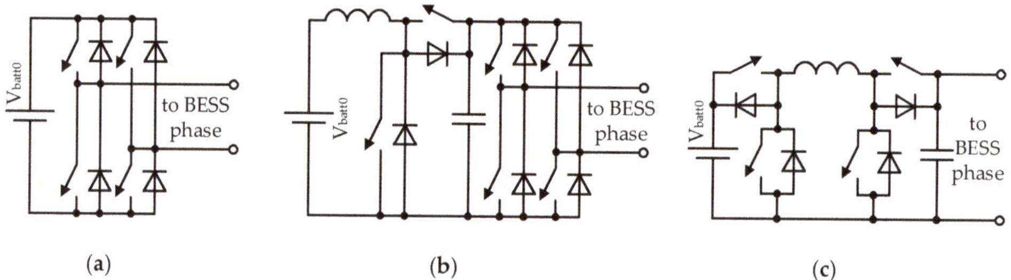

Figure 14. Power converters for multilevel converter building blocks: (**a**) full bridge or H-bridge, (**b**) H-bridge with correcting synchronous buck converter, (**c**) unipolar bidirectional converter [108].

Neutral Point Clamped Multilevel Converters

The simplest kind of the neutral point clamped multilevel converters is known as the diode clamped topology (Figure 15a). It has quite high efficiency compared to other topologies. However, there are some disadvantages: the number of power diodes is quadratic related to the level count, which makes this topology quite difficult to use when a large number of levels is needed. Another disadvantage of the topology is that charge balancing in the capacitors is needed. Another type of the neutral point clamped multilevel converter is an active clamped multilevel converter shown in Figure 15b. Additional switches enable the distribution of power losses more evenly between the switches. Besides, it is possible to synthesize 0 V level differently, providing different charge/discharge paths.

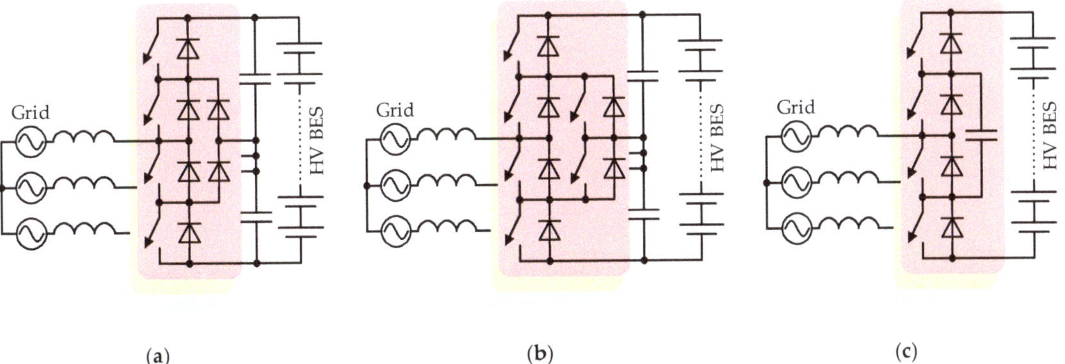

Figure 15. Multilevel converters with solid HV battery: (**a**) diode clamped, (**b**) active clamped, (**c**) flying capacitor.

The use of neutral point clamped converters in the BESS system is described in [110,111]. Reference [112] demonstrates the use of neutral point clamped and active neutral point clamped converters in BESSs. In [113], an overview of modular converters (including active neutral point clamped converters) in BESS systems is given. Diode clamped and independent source multilevel converters in BESS applications, indicating also larger operating range of the diode clamped converters, are compared in [114].

Multilevel Inverter with Flying Capacitors

The main difference between multilevel converters with neutral point clamped and multilevel converters with flying capacitors (Figure 15c) is that instead of clamping diodes, capacitors are used. Similar to the diode-clamped topology, the main disadvantage of the multilevel inverter with flying capacitors is the large number of used capacitors, which makes the practical implementation of this solution larger in terms of packaging. In spite of this drawback, some recent papers report that the topology itself can be successfully applied in BESS based on GaN switches: [115] presents a BESS with a 13-level converter, but [116]—a 9-level converter for aircraft. In addition, the BESS interface converter offered by SolarEdge in [108] utilizes the MLC with flying capacitors as a grid inverter.

4.3. *Impedance-Source Bidirectional Inverters/Rectifiers*

The problems of conventional topologies related to the battery voltage variation can be mitigated with the family of impedance source (IS) converters. These topologies incorporate a special network, which allows step-up of the input voltage using a shoot-through state in the inverter bridge, which is a prohibited condition in conventional inverters. As a result, IS converters can be less prone to short-circuit faults. There is a variety of impedance source networks proposed in the literature for a range of applications with different properties and features (Figure 16), including three-phase and multilevel configurations [117]. The majority of basic impedance source topologies were initially unidirectional; however, some studies address the bidirectional versions potentially suitable for residential BES [118,119].

4.4. *Bidirectional Two-Stage DC-AC Converters*

This section presents the state-of-the-art and emerging power electronic interfaces for BES, featuring two explicit stages. In a general case, the first stage is a bidirectional DC-DC converter, which processes varying battery voltage and controls the charge/discharge current. It operates in conjunction with the DC-AC inverter/rectifier addressed in the previous section, which provides interface with the grid.

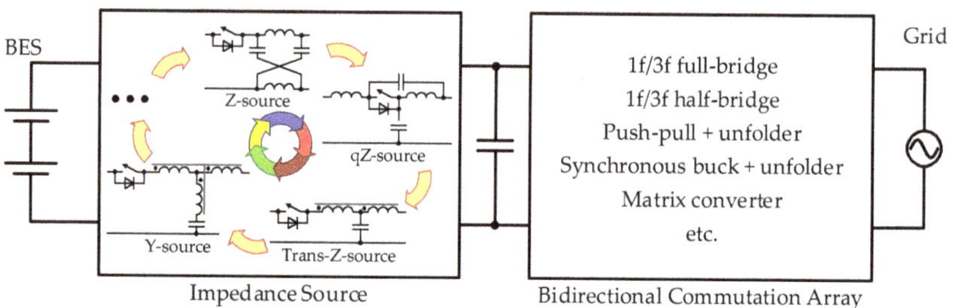

Figure 16. Generalized configuration of the BES interface converter with an impedance source.

In the most obvious operation mode, this DC-DC converter provides the stabilized voltage in the DC-link at all operation points of the battery (Figure 17a) while the rectifier-inverter modulates the voltage at the grid end according to the phase of the network voltage and required grid current. It was demonstrated that two-stage configurations are overall superior to the single DC-AC inverter/rectifier in terms of efficiency throughout the battery voltage range [86]. Moreover, the stable DC-link voltage allows integration of other DC sources and loads; therefore, such solutions can be suited for both DC- and AC-coupled BESS.

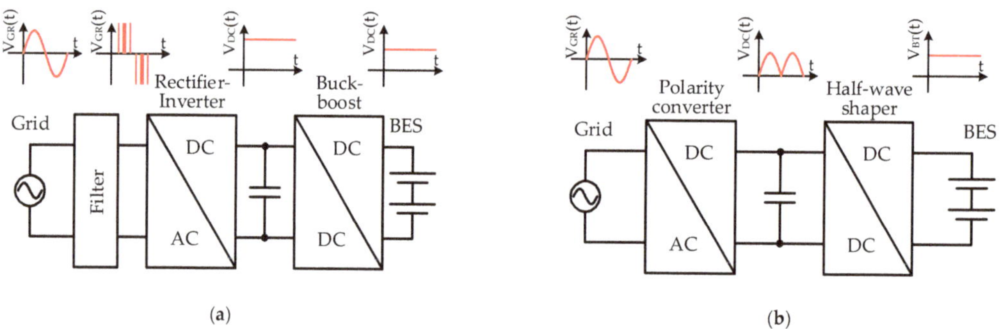

Figure 17. Operation and configuration of BES interface converters with two stages: (**a**) common DC-link, (**b**) unfolding topology.

However, one more operation mode and configuration is possible. In this mode, the DC-DC converter forms unipolar sine half-waves in the DC-link, but the rectifier-inverter applies the formed half-waves to the grid with the correct polarity (Figure 17b). In the second case, the rectifier-inverter does not operate in a real switch-mode—it just commutates the half-waves at the grid frequency. Therefore, in this operation mode, the switching losses of the rectifier-inverter are negligible, while the grid filter can be omitted or reduced due to the continuous profile of the voltage at the grid port of the rectifier-inverter [120]. In the single-phase configuration, the AC-DC converter is a bridge or half-bridge circuit, close to that shown in Figure 16 without the inductance coil. Depending on the required power and input connection, it can be a single-phase [120–122] or a three-phase [122,123] circuit.

4.4.1. Two-Stage Converters with Stabilized DC-Link

The typical configuration of a two-stage converter assumes voltage stabilization at an intermediate DC-link to compensate battery voltage variation and provide optimal operating conditions for the DC-AC inverter/rectifier. Such configuration can be suited for both DC- and AC-coupled BES. The standard DC-DC stage topologies include buck, boost, buck-boost, etc. The common disadvantage of these standard configurations is that both stages have to be rated for the full power of the system. This results in increased cost and negative impact on the efficiency. One of the recent trends in the power electronic studies is the use of advanced topologies of the DC-DC stage like differential, partial and fractional power converters that allow operation with lower voltages/currents and minimization of power losses. The use of these topologies in BES interface is considered below.

Standard Topologies of DC-DC Converters

The choice of the secondary (DC-DC) converter or the converter at the battery end depends on several factors. First of the all, this converter has to be bidirectional. Besides, this reduction of the losses requires that the number of switches is minimal, which enables only simple choppers. Finally, the configuration/mode of the two-stage interface converter (stabilizing or unfolding) as well as the voltage of the battery are important. Below, the latter issue is addressed in detail.

In the case of the two-stage converter with a stabilized DC-link, its grid unit may operate correctly if the voltage of the DC-link is higher than the amplitude of the grid voltage. On the other hand, keeping this voltage level on the battery is not reasonable because it would reduce the advantages of the two-stage configuration. Taking into account the realistic voltage gains of the circuits with the boost function of 2–3 and voltage difference of the fully discharged and fully charged Li-Ion batteries of 60–100%, the voltage of the fully charged battery could be at least twice lower than the amplitude of the grid voltage or about 200 V. In the given paper, this level is considered as a medium level, but such batteries abbreviated as MV BES. The case of the unfolding configuration/mode additionally requires that the DC-DC chopper is capable of converting the BES voltage down to zero level. Together, as a result, the following buck-boost converters are suitable for both configurations/modes: classical (Figure 18a) and non-inverting (Figure 18b) bidirectional buck-boost converter, Zeta-SEPIC circuitry that is linked through its primary inductor to the battery or the grid (Figure 18c), as well as the bidirectional Čuk converter (Figure 18d). In addition, a synchronous buck converter (Figure 18) is applicable in the case of the stabilizing configuration/mode only.

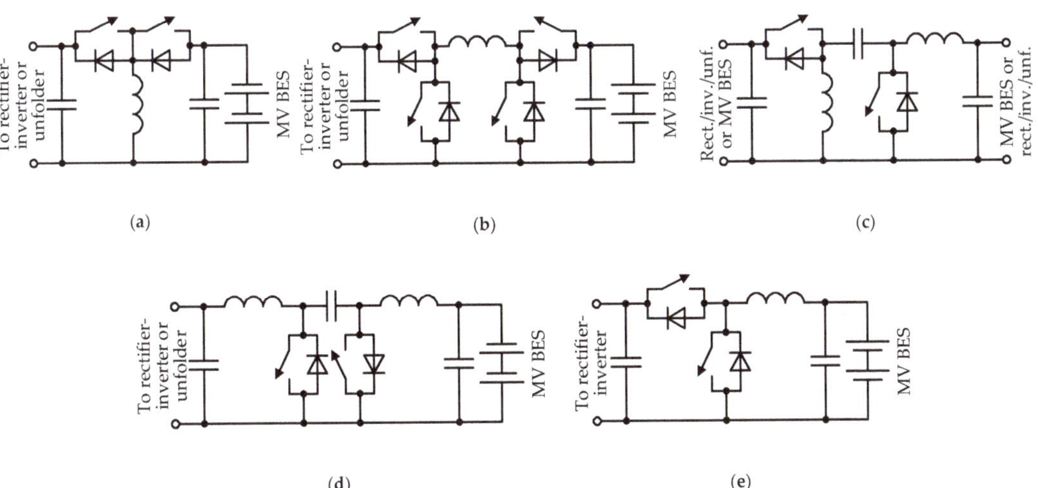

Figure 18. DC-DC stage of BES interface converters with two stages: (**a**) inverting bidirectional buck-boost, (**b**) bidirectional non-inverting buck-boost, (**c**) bidirectional zeta-SEPIC, (**d**) bidirectional Čuk, (**e**) synchronous buck.

Zeta-SEPIC-Čuk topologies, however, are usually not considered as powerful converters, which disables them for applications like BESS. On the other hand, the remaining (non)inverting buck-boost and synchronous buck topologies can be equipped with additional elements for reducing losses, smoothing current ripple and better control performance (more accurate regulation for the same range of duty cycle): add-on circuits for zero-current/resonant switching, tapped (coupled) inductors or qZ links [120].

Differential Power Converters

Differential power converters (DPCs) are a kind of partial power processor (PPP). In turn, PPPs are a quite recent group of converters that are typically used in conjunction with renewable energy sources and storages. As it follows from their title, the main feature refers to dealing only with a part of the total system power. PPPs can be systematized in a number of ways, such as considering their topology or application. However, most commonly, PPPs can be divided into different groups [124] according to their power flow. From this point of view, three groups could be differentiated: differential power converters that internally link elements of the systems, partial power converters connecting system input and output, fractional power converters dealing with a fraction of entire set of power sources/storages, as well as mixed topologies. Finally, it must be noted that the difference between differential, partial and fractional power converters is sometimes quite fuzzy. For example, in [125,126], identical topologies are entitled as partial and differential. In a similar way, most of the topologies considered in [127] as partial power converters, in fact, operate as fractional power converters.

DPCs are mainly used in various balancing systems [128,129]. There are two types of such converters. The first one transfers the energy between two typically adjacent elements and is known as element-to-element (E2E) converter; alternatively, in another option, the energy circulates through a common bus (B2E). The DPC normally operates with batteries, but according to some reports, this converter technology is applied to photovoltaics [128].

The E2E architecture is used in systems with the same type of cells, for example, for balancing batteries (Figure 19a). The main advantage of this architecture is that each converter operates at significantly lower voltage and current values than the entire system. The disadvantages of this architecture are the interconnectedness of the converters and the impossibility of their operation separately.

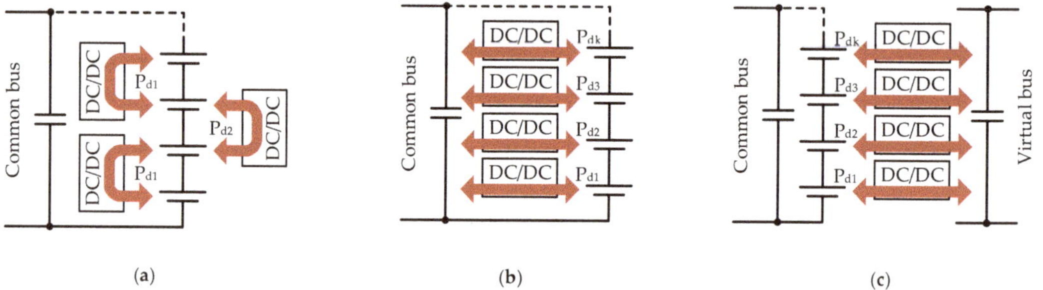

Figure 19. Differential power converters: (**a**) Element-to-Element, (**b**) Element-to-common-bus, (**c**) Element-to-virtual-bus.

The B2E architecture works with a common bus connected to the output (Figure 19b) or with the independent "virtual" bus dedicated to the energy transfer (Figure 19c). Each element is connected to the bus via own converter. Compared to the E2E architecture, this approach is more flexible, but neighboring cells are independent of each other. However, an isolating converter suitable for the full bus voltage is required.

All kinds of DPCs fit well the cell balancing function needed also in BESSs. At the same time, the use of B2E DPCs as BES and grid interface converters is complicated due to the following: (1) DC output requires an inverter or unfolder and (2) because the total power of converters, in fact, is not reduced, but just split into several parts. Lastly, the E2E DPCs are not applicable as BES and grid interface converters due to the absence of the common link.

Partial Power Converters

Another group of PPP links the input and the output of the system. While one part of the energy from the source to the load goes directly, the converter transfers only the necessary reminder. As can be seen from Figure 20, in the classical converter type (Figure 20a), all the output power passes through the converter, which leads to a higher efficiency and significant losses. In the case of a PPC, a significant part of the total energy enters the load without conversion and does not produce losses. Only the energy going through the converter, adjusted by the converter to control the energy flow, produces losses (Figure 20b). Thus, compared to a classic converter, PPCs have potentially better efficiency and smaller dimensions for the same power. The PPCs may operate with reduced voltage (Figure 20c), current or both.

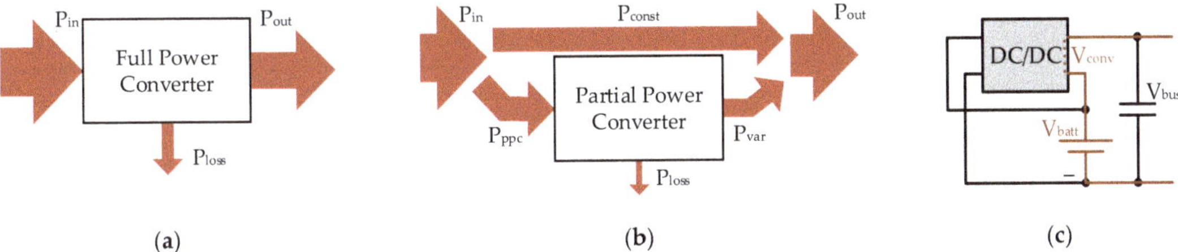

Figure 20. Full power operation vs. partial power operation: (**a**) power distribution in full power converter, (**b**) power distribution in partial power converter, (**c**) diagram showing operation with reduced voltage.

It is possible to distinguish two groups of PPCs: with an isolated and with a non-isolated converter. The isolated converter can be applied in a quite free form. That is why such PPCs can be of two types: parallel input—serial output, as well as serial input—parallel output (Figure 21). In the first case (Figure 21a), the input source and the input of the converter are connected in parallel, while the output of the converter and the input source are connected in series (S-PPC). The configuration is suitable to increase the voltage. In respect to the battery, the parallel input converters can be considered as partial current converters because only part of their battery current is transferred to the output (bus) through the converter. On the other hand, in respect to the output, operation occurs with reduced voltage because only part of output voltage is applied to the converter (see also Figure 20c).

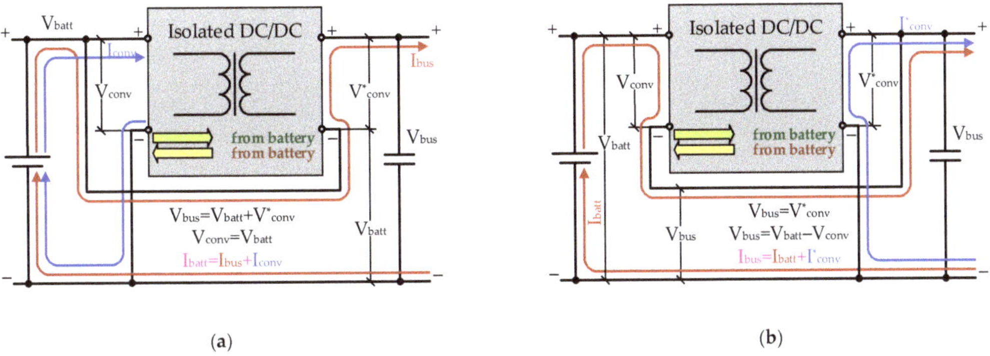

Figure 21. Structures of partial power converters: (**a**) parallel input—series output, (**b**) parallel output—series input.

In the second case (Figure 21b), the input source is connected in series with the input of the converter, but its output and the source are connected in parallel (P-PPC). The configuration is suitable to obtain higher output current. In respect to the battery, the series input converters can be considered as partial voltage converters because only part of their battery voltage is converted and passed to the output.

It is obvious that both topologies are symmetrical and counter-reversible. In respect to the output (bus), the first configuration is a partial voltage converter, but the second one—a partial current converter. To some extent, these PPCs are similar to an autotransformer and can be described by similar mathematical expressions extracted from Kirchhoff's voltage and current laws.

The PPC topology provides benefits when the difference between the input and the output voltage is relatively small and only a small amount of energy is being converted by PPC. Due to a more complex design and a larger number of active elements, the larger difference between the input and the output voltages produces lower efficiency. Moreover, at 100% of the difference, the efficiency will be less than that of a classical converter.

Practical PPC implementation depends on the particular application. Normally, reports consider PPC with a DAB converter at each end of an isolating transformer that produces a fully bidirectional PPC (Figure 22a). In many applications, the bidirectionality can be omitted, but PPC—reasonably simplified. For example, in [130,131], which address PV systems, the simplification finally produces full-bridge + buck configuration, in [131,132]—full-bridge + push-pull, but in [130,132]—a kind of classical flyback. The latest converter can be easily turned to a bidirectional one (Figure 22b).

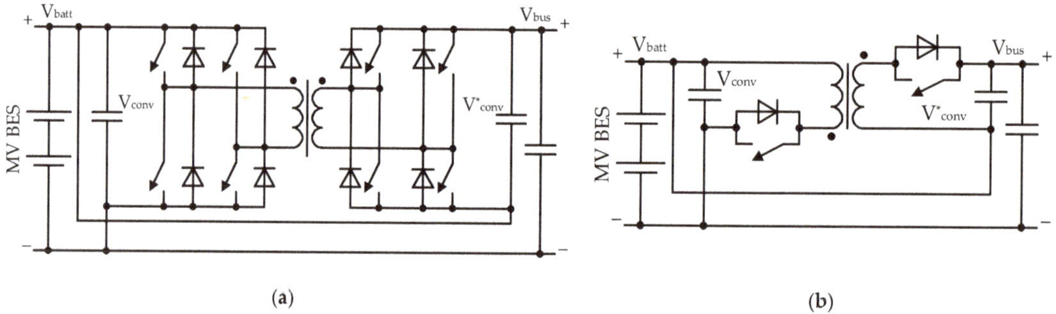

Figure 22. Examples of PPCs: (**a**) parallel input—series output PPC with DAB, (**b**) parallel input—series output PPC with flyback converter.

A PPC topology with a non-isolating converter could be potentially simpler, contain fewer components, and have higher efficiency. Attempts have been made to implement such non-isolating schemes. For example, refs. [133–135] report the voltage buck-boost topology based implementation applied for battery or bus voltage magnification. It is pointed out that the extra feedback capacitor installed in these schematics is required for direct power feedforwarding. However, it is possible to show that the obtained converters are, in fact, ordinary boost or buck converters—see [136] for details.

Fractional Power Converters

Fractional Power Converters (FPCs) deal with an explicit part of the entire power supply, for example, with several cells of BES [137,138]. In contrast to PPCs, where the reduced operating voltage of the converter is obtained as a difference on the entire input/output, FPCs process already reduced voltage—a section of the entire power supply (similar consideration could be applied to current conversion).

The fractional power processing may utilize an isolated (Figure 23b) or a non-isolated (Figure 23c) scheme. Successful examples of non-isolated converter use have been demonstrated in conjunction with a battery of PEVs [137] or grid BESS [138]. The mentioned reports explore several DC-DC choppers functioning as FPC (Figure 24 shows discharging configurations in black, but charging—in gray). In this case, the non-isolated converter obviously deals with reduced voltage, thus providing true partial power processing. On the other hand, the fraction of the power supply associated with the converter operates differently from the rest of BES. It has different average charge/discharge current. Moreover, depending on the applied chopper, it may conduct pulse-mode current. This may lead to shorter operation cycles, limited state of charge and, finally, may lead to a worse state of health for the "processed" cells.

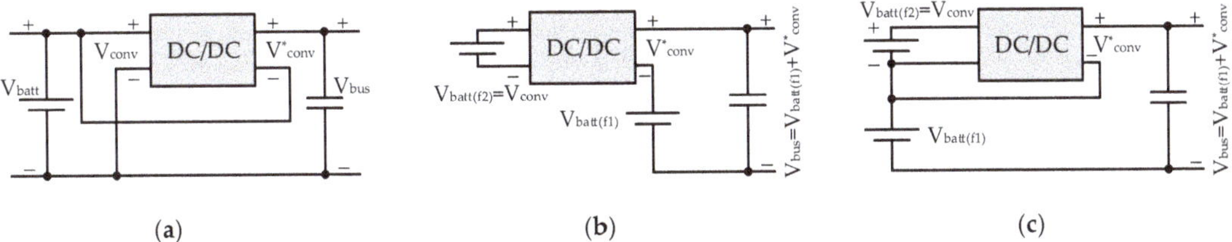

Figure 23. Partial power converters (**a**) vs. fractional power converters: (**b**) isolated, (**c**) non-isolated.

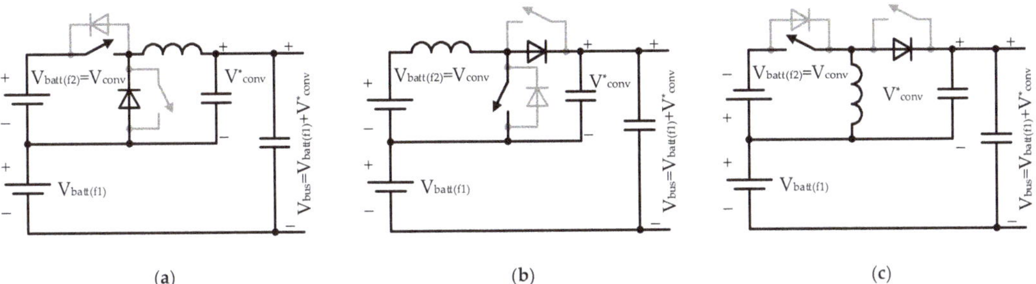

Figure 24. Examples of non-isolated fractional power converters (in discharge mode): (**a**) buck, (**b**) boost, (**c**) buck-boost.

The FPC shown in Figure 24c is also quite impractical because the polarities of the input/output voltage are different, which splits the battery or narrows the regulation range. On the other hand, the use of non-inverting buck-boost topology would double the static and dynamic losses of the switches.

Finally, the considerations on the non-isolated PPC with a feedback capacitor (given in the previous section) may also produce, in fact, an FPC if the feedback capacitor is substituted with an energy source or storage (battery, supercapacitor, PV cells etc.) capable of keeping its voltage at a constant level. Then the part of the current is actually bypassed, but the other—processed in the converter (Figure 25a). Practical importance of this converter is questionable because one fraction of the battery is loaded with increased current and charged/discharged more intensively.

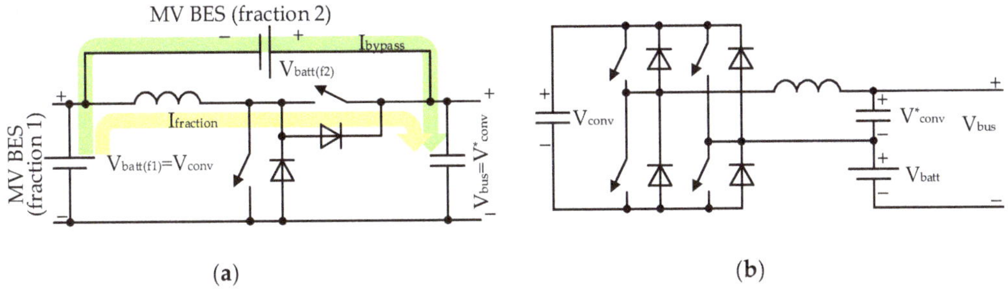

Figure 25. Specific kinds of FPC: (**a**) FPC synthesized from "non-isolated" synchronous buck PPC, (**b**) fractional power conversion with "virtual fraction" [139].

Another example of the implementation of non-isolated partial power conversion is given in [139]. This example, however, may also be considered as a fractional power converter with "virtual" fraction formed of a DAB and an ordinary capacitor. Here, the DAB processes low voltage of the capacitor while the battery is attached in series without processing (Figure 25b). Although, in [139], with the focus on the DC-AC systems, partial conversion of power occurs in the DC-link. Due to the limited energy capacity, the configuration is suitable for compensation of regular short term voltage fluctuation or current compensation that happens, for example, within the cycle of the supply grid (20 ms).

4.4.2. Two-Stage Converters with Pulsating DC-Link

In order to reduce the overall switching losses of the two-stage system, a configuration with inverting unfolder can be used. In this case, the DC-DC converter forms unipolar sine half-waves in the pulsating DC-link, but the interface inverter applies it to the grid with the proper polarity.

The Li-ion battery can handle the current ripple without significant effect on their lifetime, thus the use of pulsating current can be justified [140,141].

1-ph Unfolders

As it was stated previously, an unfolding circuit provides grid-frequency commutation of the unipolar voltage formed by a high-frequency switch mode DC-DC converter to provide sinewave matching to the grid polarity. Paper [142] proposes a combination of a buck/boost non-inverting converter and an unfolding H-bridge (Figure 26a). This configuration directly corresponds to Figure 17b and can be considered as a standard double stage converter with a pulsating DC-link. In [143], the operation and the experimental verification of a buck-boost inverter/converter based on tapped inductor are addressed. The inductor magnetically couples four windings with equal turn-ratio (Figure 26b). In the converter presented in [143], in contrast to [142], explicit parts cannot be identified.

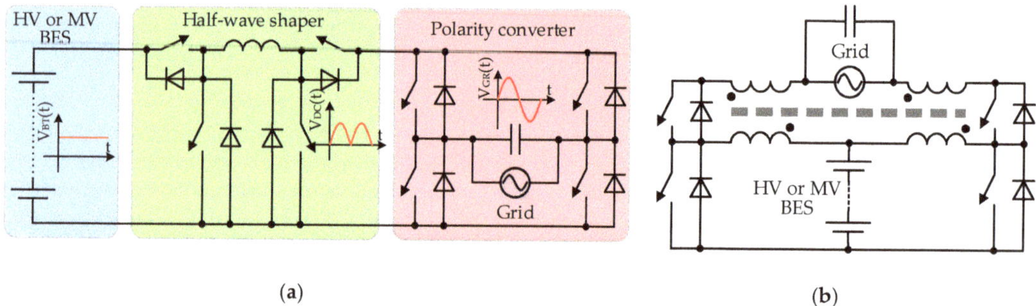

Figure 26. Examples of single-phase unfolders: (**a**) buck-boost converter with unfolding H-bridge, (**b**) tapped inductor buck-boost converter.

3-ph Unfolders

Three-phase converters with low frequency unfolding stage utilize principles similar to those of single-phase unfolders. However, the presence of three phases requires that at least two voltages/currents be formed actively by the dedicated voltage/current source (shaper), but the third one is obtained as a sum/difference of the other two. Within a period of the grid, the principles how the actively shaped voltages/currents are applied to the grid change six times (Figure 27a,b).

Working principles of unfolder topologies are provided in [144]; however, the converters described there are unidirectional and do not fit the requirements of bidirectional operation. An example of such converter from [144] is a topology derived from a three-phase two-level voltage source inverter. In this case, amplitude modulated high-frequency output of a phase modulated high-frequency inverter (H-bridge was taken as an example) is rectified and filtered. Then the output of a filter is unfolded by a three-phase inverter, thus forming three-phase alternating current. For the bidirectional operation, one or multiple DC-DC converters should be used as current sources. For example, refs. [145,146] show a three-phase inverter, where two DC-DC converters were taken as current sources and are connected in parallel to the BES. DC-DC converter outputs are connected in series, thus forming three voltage levels—high, low and neutral. Then the modulated voltage waveforms are unfolded by a three-level inverter, which is derived from a diode clamped multilevel converter [146] or Vienna rectifier [145] (Figure 27c). The study in [147] provides experimental verification of the topology in [146].

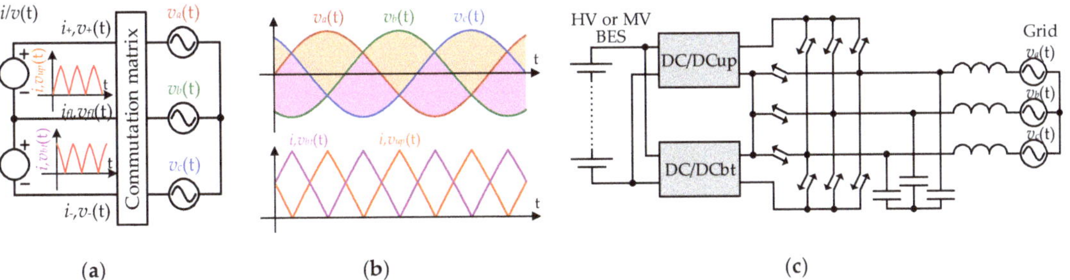

Figure 27. Three-phase unfolders: (**a**) general principle, (**b**) operation diagram, (**c**) example of implementation.

5. Generalizations and Discussion

One of the main trends in the area of residential BESSs is a shift from stand-alone systems for a particular task, like the increasing local renewable energy self-consumption using basic algorithms, to a complex integrated system, aiming to achieve multiple different goals. Some commercial BESSs can provide a supply during a power cut, effectively acting as uninterruptable power supplies. The use of advanced energy management algorithms enables smart scheduling and energy trading, further improving the functionality of these systems. The integration on a system level allows provision of ancillary grid services and enables forming of a microgrid with extended levels of flexibility. Nevertheless, currently there are still a number of technological and legal barriers that limit a ubiquitous application of the residential BESS. However, the new regulations, subsidies and initiatives offered and developed by governmental structures in many countries are aimed to stimulate the installation of these systems and make full use of their potential [148].

Our review of the commercially available BESS and the corresponding intellectual property right items shows that despite a whole range of available solutions, the market of residential BESSs is still advancing. Presently, it is strongly influenced by a forecasted price reduction of the Li-ion battery cells and further improvements in the battery chemistry. Technical information about available products is rather limited. Most of the BESSs utilize a low-voltage (around 50 V) Li-ion battery, which results in high current and requires the use of a transformer. Most likely, two-stage converters with galvanic isolation are used, limiting the overall efficiency of the system. The typical BESSs have the efficiency below 90%, whereas power electronic converters could be responsible for around half of power losses in the system. On the other hand, the low voltage batteries require simpler battery management and protection systems, making them less expensive and more feasible commercially. Because of these compromises, the current BESSs have strict limitations on the efficiency due to the use of low voltage batteries that are associated with high currents and more complex converter topologies. On the other hand, the batteries with increased voltages would enable the use of non-isolated topologies with potentially higher efficiencies. Several companies have already introduced such products to the market.

The typical power electronic interface of a battery with the grid is based on a two-stage configuration, comprised of a bidirectional DC-DC converter and a DC-AC inverter/rectifier connected via an intermediate DC-link. Modern Li-ion batteries can sustain current ripples associated with the grid frequency very well, even in single-phase systems. It is therefore possible to connect a battery of sufficiently high voltage to an inverter directly. Still, due to variation of battery voltage depending on its state of charge, the efficiency and power quality of such system is compromised. As a result, the intermediate DC-DC stage is still necessary to stabilize the DC voltage and obtain better performance.

With the possibility of using non-isolated converters to interface HV batteries, the standard approach would assume application of well-known DC-DC topologies, like buck, boost and buck-boost together with the grid inverter stage. This makes both of the two conversion stages process full power and exhibit high switching frequency, which still compromises the efficiency. One of the approaches that is widely addressed in recent studies is to use emerging solutions, like PPCs and FPCs at the DC-DC stage, which, as it is already reported, have been achieved extremely high efficiency values. However, the practical aspects, including transient operating modes, protection and cost, need to be evaluated further to justify this technology.

A range of alternative concepts utilizes a pulsating DC-link instead of the stabilized one. This brings the converter system closer to the single-stage converter, where only DC-DC stage operates with high frequency, while the grid-side inverter just unfolds the unipolar pulses into the sine wave and exhibits conduction losses only. A similar approach can be applicable to both single- and three-phase systems [145]. In addition, mixed concepts with fluctuating DC-link were also proposed, aiming to distribute losses more evenly between the stages [149]. On the other hand, it would be much more difficult to integrate other sources into such DC-link and therefore such solutions are generally suitable for AC-coupled BESS only.

Impedance source inverters are another group of topologies that allow voltage pre-regulation at a "virtual DC-link" before it is inverted into a sinusoidal waveform. Single-phase, three-phase and three-level configurations of these inverters were proposed in [150]. They can be more short-circuit-proof, as the shoot-through state is one of the inherent operating modes of such topologies. However, some studies show that the voltage stress on semiconductors and volume of components can be larger than for the standard two-stage configurations [151]. Moreover, only few studies address bidirectional operation of impedance-source converters [152,153].

In conclusion, there is a range of solutions for HV BESSs that are potential alternatives to standard buck-boost plus inverter configuration. The most optimal choice would evidently depend on the parameters of the system and its configuration. For the systems that incorporate a DC bus for integration of renewables and loads, a PPC/FPC with a bridge-type bidirectional inverter/rectifier seems to be a very promising solution. On the other hand, for an AC-coupled BES, the use of pulsating/fluctuating DC-link and unfolding inverter can bring an advantage in terms of switching loss and absence of a bulky capacitor. Still, the behavior of such configurations in practical applications, including transient modes and fault ride-through capabilities, needs to be addressed in more detail.

The configurations that include multilevel inverter topologies also seem quite promising for residential BESs. Despite generally being used in high-power applications, there are successful commercialization examples of this technology in residential applications. Recent works aim to bring such inverters on a new level, particularly taking advantage of developments in WBG semiconductors [115,116]. The systems with multilevel topologies potentially enable the use of battery stacks with lower voltage levels as compared to standard two-level inverters. This could result in a more optimal storage configuration. On the other hand, presently, the cost of WBG devices is still relatively high to make such multilevel inverters feasible in commercial BESSs.

The comparative analysis of evaluated power electronic interfaces is presented in Tables 3 and 4. The considerations above show that the most promising units of composite

BESS grid interface converters have somehow completing features (see Table 3 for details). For example, unfolding circuits provide neither DC regulation at the corresponding port nor AC half wave forming. This functionality, however, can be performed by a DC-DC converter. Multilevel converters without pulse mode control do not provide pulse mode regulation between levels, but partial power converters—provide regulation within a narrow range. Besides, the multilevel converters and unfolding units have no switching losses, but have significant conduction losses (Table 3). At the same time, the partial power converters can reduce both. A logical conclusion from the above is to combine the units with the adjacent features (Table 4).

Table 3. Losses of converters and energy conversion principles in BESS grid interface.

Stage	Main Function	Peculiarities
Full Power Switch-Mode Rectifiers/Inverter (origin for comparison)	Forming AC	+ Established technology, − High voltage input, high switching frequency, bulky filter
Full Power Switch-Mode DC/DC Converters	Forming DC	+ Established technology, wide regulation range, − Full power operation, high switching frequency
Partial Power Converters	Forming DC	+ Operation with part of rated power − Developing technology, limited regulation range
Multilevel Converters	Forming AC or DC	+ Established technology, small grid filter − Control and hardware complexity
Unfolding Circuits	Commutation	+ No switching losses − Developing technology, no regulation

Table 4. Promising combination of converters to form BESS grid interface.

Configuration	Advantages	Disadvantages
Single stage DC-AC Bidirectional Inverters/Rectifiers	Max. efficiency at a particular operation point	Lower efficiency at most of the operation points, Minimal battery voltage > amplitude of grid voltage
Impedance-Source Bidirectional Inverters/Rectifiers	Battery voltage pre-regulation Short-Circuit Proof	Voltage stress on semiconductors and volume of components is larger Complicated bidirectional operation Developing technology
Bidirectional inverter/ rectifier + Full Power DC-DC	Higher efficiency at the most of operation points, Wide battery voltage range, Allows integration of renewables into DC-link	Lower maximal efficiency, Both stages operate at full power and high switching frequency
Bidirectional inverter/ rectifier + PPC DC-DC	Higher efficiency at the most of operation points, Allows integration of renewables into DC-link, DC-DC operates with part of rated power	Narrow battery voltage range, Developing technology
Multilevel DC-DC and DC-AC	Low grid filter size and volume, Utilization of low voltage semiconductors, Modular design	High component count Complex control
Unfolder + Full Power DC-DC	Higher efficiency at the most of operation points, Wide battery voltage range, No switching losses in grid stage, No DC-link capacitor	No integration of renewables into DC-link

6. Conclusions and Future Trends

This paper gives an insight into the field of storage systems for residential applications together with associated technologies and developments. To provide a broader view, the current state of these systems is addressed from multiple directions, including battery technologies, their market, standards, and grid interface converters.

Instigated by the on-going paradigm shift from centralized to distributed power generation, the storage technologies will become one of the key components of the future electrical grids that enable more optimal use of the conventional and local renewable energy sources and ensure the power supply security. However, a range of technological and regulatory barriers still stand in the way of these systems, limiting their benefits and potential.

Today's market for dedicated residential storage systems is still in the process of being established. It is currently very dynamic, and several manufacturers have already introduced and commercialized their solutions, with more companies and products being announced and trying to enter the market every year. Still, the price for residential solutions is relatively high for a private client, while the return of investment is not evident in many cases.

The developments and price reduction of Li-ion battery technologies are mainly driven by massive transportation electrification and this trend will continue in the following years. Despite the distinct potential of vehicle to grid (V2G) solutions, they are unlikely to be able to replace stationary battery systems and their functions due to economic reasons, mainly related to lifetime and cycle-cost. Nevertheless, the use of second-life Li-ion batteries for stationary storage has certain potential.

Batteries based on the Li-ion technology are currently dominating the market, however, at a certain point, the price and performance of other battery technologies, like flow batteries, is likely to make them a more expedient choice for larger-scale stationary solutions.

According to our analysis, the majority of commercial residential storages are currently using low voltage batteries with voltages of around 50 V, mainly due to the cheaper price per kWh. These batteries are typically interfaced with the grid by means of a power electronic converter with a transformer to provide required voltage matching and galvanic isolation. However, the mass production of HV EV batteries along with their second-life use is likely to make the HV stationary storage solutions more popular in the residential sector. This would make the use of non-isolated interface converter topologies attractive due to their typically lower component count and higher efficiency. In addition to standard and typically used topologies, like buck-boost or bridge, which are rated for full power of the system, the recent research interest is also focused on partial- (fractional-, differential-) power converters. Such topologies have the potential to offer even further improvement of efficiency in various operating conditions.

Presently, many countries are introducing initiatives that are either directly (by subsidies) or indirectly (via marginal feed-in tariffs) encouraging the use of local energy storage. Moreover, a range of standards is being developed to regulate the use of such systems and facilitate unleashing of their full potential. In addition to basic renewable energy self-consumption increase, battery-based storage systems can provide uninterruptable power supply functionality, offer ancillary grid service support, enable peer-to-peer energy trading etc. Together with the large-scale global investments in the battery technologies it is highly likely that in the following decades, the residential battery systems will follow the route of photovoltaics and become an essential and inherent part of the future power grid.

Author Contributions: Conceptualization, I.A.G. and A.B. (Andrei Blinov); investigation of BESS market solutions, R.S.; investigation of IP right items and BESS market state for distribution grids, M.V.; topological analysis of single stage pulse mode converters, I.A.G.; topological analysis of multilevel converters, A.B. (Alexander Bubovich); topological analysis of partial power processors, I.A.G. and R.S.; supervision, I.A.G.; Validation, I.A.G.; writing—original draft, all; Supervision and editing—A.B. (Andrei Blinov) and D.P. All authors have read and agreed to the published version of the manuscript.

Funding: This work was supported in part by the European Economic Area (EEA) and Norway Financial Mechanism 2014–2021 under Grant EMP474 and in part by the Estonian Research Council grant (PRG1086).

Institutional Review Board Statement: Not applicable.

Informed Consent Statement: Not applicable.

Data Availability Statement: Data is contained within the article.

Conflicts of Interest: The authors declare no conflict of interest.

References

1. The European Parliament and Council of the European Union. Directive 2009/28/EC on the promotion of the use of energy from renewable sources. *Off. J. Eur. Union* **2009**, *L140*, 16–62.
2. The European Parliament and Council of the European Union. Directive (EU) 2018/2001 on the promotion of the use of energy from renewable sources. *Off. J. Eur. Union* **2018**, *L328*, 82–209.
3. European Commission. Energy Strategy. Available online: https://ec.europa.eu/energy/topics/energy-strategy-and-energy-union_en (accessed on 13 November 2020).
4. Andrijanovits, A.; Hoimoja, H.; Vinnikov, D. Comparative Review of Long-Term Energy Storage Technologies for Renewable Energy Systems. *Elektron. Elektrotechnika* **2012**, *118*, 21–26. [CrossRef]
5. Vinnikov, D.; Hoimoja, H.; Andrijanovits, A.; Roasto, I.; Lehtla, T.; Klytta, M. An improved interface converter for a medium-power wind-hydrogen system. In Proceedings of the 2009 International Conference on Clean Electrical Power, Capri, Italy, 9–11 June 2009; Institute of Electrical and Electronics Engineers (IEEE): New York, NY, USA, 2009; pp. 426–432. [CrossRef]
6. Liang, M.; Liu, Y.; Xiao, B.; Yang, S.; Wang, Z.; Han, H. An analytical model for the transverse permeability of gas diffusion layer with electrical double layer effects in proton exchange membrane fuel cells. *Int. J. Hydrog. Energy* **2018**, *43*, 17880–17888. [CrossRef]
7. Liang, M.; Fu, C.; Xiao, B.; Luo, L.; Wang, Z. A fractal study for the effective electrolyte diffusion through charged porous media. *Int. J. Heat Mass Transf.* **2019**, *137*, 365–371. [CrossRef]
8. Vinnikov, D.; Andrijanovits, A.; Roasto, I.; Jalakas, T. Experimental study of new integrated DC/DC converter for hydrogen-based energy storage. In Proceedings of the 2011 10th International Conference on Environment and Electrical Engineering, Rome, Italy, 8–11 May 2011; Institute of Electrical and Electronics Engineers (IEEE): New York, NY, USA, 2011; pp. 1–4.
9. Stecca, M.; Elizondo, L.R.; Soeiro, T.; Bauer, P.; Palensky, P. A Comprehensive Review of the Integration of Battery Energy Storage Systems into Distribution Networks. *IEEE Open J. Ind. Electron. Soc.* **2020**, *1*, 46–65. [CrossRef]
10. Wang, G.; Konstantinou, G.; Townsend, C.D.; Pou, J.; Vazquez, S.; Demetriades, G.D.; Agelidis, V.G. A Review of Power Electronics for Grid Connection of Utility-Scale Battery Energy Storage Systems. *IEEE Trans. Sustain. Energy* **2016**, *7*, 1778–1790. [CrossRef]
11. Pires, V.F.; Romero-Cadaval, E.; Vinnikov, D.; Roasto, I.; Martins, J. Power converter interfaces for electrochemical energy storage systems–A review. *Energy Convers. Manag.* **2014**, *86*, 453–475. [CrossRef]
12. Yao, Z. Review of Dual-Buck Type Single-Phase Grid-Connected Inverters. *IEEE J. Emerg. Sel. Top. Power Electron.* **2020**, *1*. [CrossRef]
13. Available online: https://nakedsolar.co.uk/storage/ (accessed on 7 June 2021).
14. Available online: https://www.which.co.uk/reviews/solar-panels/article/solar-panels/solar-panel-battery-storage-a2AfJ0s5tCyT (accessed on 7 June 2021).
15. Saez-De-Ibarra, A.; Laserna, E.M.; Stroe, D.-I.; Swierczynski, M.J.; Rodriguez, P. Sizing Study of Second Life Li-ion Batteries for Enhancing Renewable Energy Grid Integration. *IEEE Trans. Ind. Appl.* **2016**, *52*, 4999–5008. [CrossRef]
16. Rezania, R.; Prüggler, W. Business models for the integration of electric vehicles into the Austrian energy system. In *2012 9th International Conference on the European Energy Market*; Institute of Electrical and Electronics Engineers (IEEE): New York, NY, USA, 2012; pp. 1–8.
17. Robert, S. *Introduction to Batteries—An IEEE Course*; IEEE: New York, NY, USA, 2013.
18. Hu, X.; Zou, C.; Zhang, C.; Li, Y. Technological Developments in Batteries: A Survey of Principal Roles, Types, and Management Needs. *IEEE Power Energy Mag.* **2017**, *15*, 20–31. [CrossRef]
19. *Electricity Storage and Renewables: Costs and Markets to 2030*; The International Renewable Energy Agency (IRENA): Abu Dhabi, United Arab Emirates, 2017; pp. 36–49.
20. Cao, J.; Emadi, A. Batteries Need Electronics. *IEEE Ind. Electron. Mag.* **2011**, *5*, 27–35. [CrossRef]
21. Ferreira, B. Batteries, the New Kids on the Block. *IEEE Power Electron. Mag.* **2019**, *6*, 32–34. [CrossRef]
22. Chen, C.; Plunkett, S.; Salameh, M.; Stoyanov, S.; Al-Hallaj, S.; Krishnamurthy, M. Enhancing the Fast Charging Capability of High-Energy-Density Lithium-Ion Batteries: A Pack Design Perspective. *IEEE Electrif. Mag.* **2020**, *8*, 62–69. [CrossRef]
23. Fulli, G.; Masera, M.; Spisto, A.; Vitiello, S. A Change is Coming: How Regulation and Innovation Are Reshaping the European Union's Electricity Markets. *IEEE Power Energy Mag.* **2019**, *17*, 53–66. [CrossRef]
24. Lukic, S.; Emadi, A. Charging ahead. *IEEE Ind. Electron. Mag.* **2008**, *2*, 22–31. [CrossRef]
25. Khaligh, A.; Li, Z. Battery, Ultracapacitor, Fuel Cell, and Hybrid Energy Storage Systems for Electric, Hybrid Electric, Fuel Cell, and Plug-In Hybrid Electric Vehicles: State of the Art. *IEEE Trans. Veh. Technol.* **2010**, *59*, 2806–2814. [CrossRef]
26. Quiros-Tortos, J.; Ochoa, L.; Butler, T. How Electric Vehicles and the Grid Work Together: Lessons Learned from One of the Largest Electric Vehicle Trials in the World. *IEEE Power Energy Mag.* **2018**, *16*, 64–76. [CrossRef]

27. Chen, N.; Ma, J.; Li, M.; Wang, M.; Shen, X.S. Energy Management Framework for Mobile Vehicular Electric Storage. *IEEE Netw.* **2019**, *33*, 148–155. [CrossRef]
28. Chandler, S.; Gartner, J.; Jones, D. Integrating Electric Vehicles with Energy Storage and Grids: New Technology and Specific Capabilities Spur Numerous Applications. *IEEE Electrif. Mag.* **2018**, *6*, 38–43. [CrossRef]
29. Al-Rubaye, S.; Al-Dulaimi, A.; Ni, Q. Power Interchange Analysis for Reliable Vehicle-to-Grid Connectivity. *IEEE Commun. Mag.* **2019**, *57*, 105–111. [CrossRef]
30. Arboleya, P.; Bidaguren, P.; Armendariz, U. Energy Is on Board: Energy Storage and Other Alternatives in Modern Light Railways. *IEEE Electrif. Mag.* **2016**, *4*, 30–41. [CrossRef]
31. Sinhuber, P.; Rohlfs, W.; Sauer, D.U. Study on power and energy demand for sizing the energy storage systems for electrified local public transport buses. In *2012 IEEE Vehicle Power and Propulsion Conference*; Institute of Electrical and Electronics Engineers (IEEE): New York, NY, USA, 2012; pp. 315–320. [CrossRef]
32. Guarnieri, M.; Morandin, M.; Ferrari, A.; Campostrini, P.; Bolognani, S. Electrifying Water Buses: A Case Study on Diesel-to-Electric Conversion in Venice. *IEEE Ind. Appl. Mag.* **2017**, *24*, 71–83. [CrossRef]
33. Sorensen, A.J.; Skjetne, R.; Bo, T.; Miyazaki, M.R.; Johansen, T.A.; Utne, I.B.; Pedersen, E. Toward Safer, Smarter, and Greener Ships: Using Hybrid Marine Power Plants. *IEEE Electrif. Mag.* **2017**, *5*, 68–73. [CrossRef]
34. Paul, D. A History of Electric Ship Propulsion Systems [History]. *IEEE Ind. Appl. Mag.* **2020**, *26*, 9–19. [CrossRef]
35. Vicenzutti, A.; Bosich, D.; Giadrossi, G.; Sulligoi, G. The Role of Voltage Controls in Modern All-Electric Ships: Toward the all electric ship. *IEEE Electrif. Mag.* **2015**, *3*, 49–65. [CrossRef]
36. Roboam, X.; Sareni, B.; De Andrade, A. More Electricity in the Air: Toward Optimized Electrical Networks Embedded in More-Electrical Aircraft. *IEEE Ind. Electron. Mag.* **2012**, *6*, 6–17. [CrossRef]
37. Misra, A. Energy Storage for Electrified Aircraft: The Need for Better Batteries, Fuel Cells, and Supercapacitors. *IEEE Electrif. Mag.* **2018**, *6*, 54–61. [CrossRef]
38. Crittenden, M. Ultralight batteries for electric airplanes. *IEEE Spectr.* **2020**, *57*, 44–49. [CrossRef]
39. Manz, D.; Piwko, R.; Miller, N. Look before You Leap: The Role of Energy Storage in the Grid. *IEEE Power Energy Mag.* **2012**, *10*, 75–84. [CrossRef]
40. Farrokhabadi, M.; Solanki, B.V.; Canizares, C.A.; Bhattacharya, K.; Koenig, S.; Sauter, P.S.; Leibfried, T.; Hohmann, S. Energy Storage in Microgrids: Compensating for Generation and Demand Fluctuations While Providing Ancillary Services. *IEEE Power Energy Mag.* **2017**, *15*, 81–91. [CrossRef]
41. Cagnano, A.; De Tuglie, E.; Mancarella, P. Microgrids: Overview and guidelines for practicalimplementations and operation. *Appl. Energy* **2020**, *258*, 114039. [CrossRef]
42. Torres-Moreno, J.L.; Gimenez-Fernandez, A.; Perez-Garcia, M.; Rodriguez, F. EnergyManagement Strategy for Micro-Grids with PV-Battery Systems and Electric Vehicles. *Energies* **2018**, *11*, 522. [CrossRef]
43. Lezynski, P.; Szczesniak, P.; Waskowicz, B.; Smolenski, R.; Drozdz, W. Design andimplementation of a fully controllable cyber-physical system for testing energy storage systems. *IEEE Access* **2019**, *7*, 47259–47272. [CrossRef]
44. Restrepo, C.; Salazar, A.; Schweizer, H.; Ginart, A. Residential Battery Storage: Is the Timing Right? *IEEE Electrif. Mag.* **2015**, *3*, 14–21. [CrossRef]
45. González, I.; Calderón, A.J.; Portalo, J.M. Innovative Multi-Layered Architecture forHeterogeneous Automation and Moni-toring Systems: Application Case of a Photovoltaic SmartMicrogrid. *Sustainability* **2021**, *13*, 2234. [CrossRef]
46. James, G.; Peng, W.; Deng, K. Managing Household Wind-Energy Generation. *IEEE Intell. Syst.* **2008**, *23*, 9–12. [CrossRef]
47. Duryea, S.; Islam, S.; Lawrance, W. A battery management system for stand-alone photovoltaic energy systems. *IEEE Ind. Appl. Mag.* **2001**, *7*, 67–72. [CrossRef]
48. Lu, X.; Wang, J. A Game Changer: Electrifying Remote Communities by Using Isolated Microgrids. *IEEE Electrif. Mag.* **2017**, *5*, 56–63. [CrossRef]
49. Zhong, Q.-C.; Wang, Y.; Ren, B. Connecting the Home Grid to the Public Grid: Field Demonstration of Virtual Synchronous Machines. *IEEE Power Electron. Mag.* **2019**, *6*, 41–49. [CrossRef]
50. Ma, Z.; Pesaran, A.; Gevorgian, V.; Gwinner, D.; Kramer, W. Energy Storage, Renewable Power Generation, and the Grid: NREL Capabilities Help to Develop and Test Energy-Storage Technologies. *IEEE Electrif. Mag.* **2015**, *3*, 30–40. [CrossRef]
51. Rodriguez-Diaz, E.; Chen, F.; Vasquez, J.C.; Guerrero, J.M.; Burgos, R.; Boroyevich, D. Voltage-Level Selection of Future Two-Level LVdc Distribution Grids: A Compromise Between Grid Compatibiliy, Safety, and Efficiency. *IEEE Electrif. Mag.* **2016**, *4*, 20–28. [CrossRef]
52. IEC TC 120 Group BESS Systems Standardization Plan. Available online: https://assets.iec.ch/public/miscfiles/sbp/120.pdf (accessed on 28 February 2021).
53. *UL STD 1741. Inverters Converters and Controllers for Use in Independent Power Systems*; IEEE: New York, NY, USA, 2018.
54. *IEC TS 62933-5-1. Electrical Energy Storage (EES) Systems-Part 5-1: Safety Considerations for Grid-Integrated EES Systems-General Specification*; IEEE: New York, NY, USA; ISO/IEC: Geneva, Switzerland, 2017.
55. *IEC T R 62543. High-Voltage Direct Current (HVDC) Power Transmission Using Voltage Sourced Converters (VSC)*; ISO/IEC: Geneva, Switzerland, 2017.
56. *Standard IEC TR 61850-90-7. Communication Networks and Systems for Power Utility Automation-Part 90-7: Object Models for Power Converters in Distributed Energy Resources (DER) Systems*; ISO/IEC: Geneva, Switzerland, 2013.

57. *Standard IEC 62920. Photovoltaic Power Generating Systems-EMC Requirements and Test Methods for Power Conversion Equipment*; ISO/IEC: Geneva, Switzerland, 2017.
58. *Standard IEC 62909-1. Bi-Directional Grid Connected Power Converters-Part 1: General Requirements*; ISO/IEC: Geneva, Switzerland, 2017.
59. *Standard IEC 62040-5-3. Uninterruptible Power Systems (UPS)-Part 5-3: DC Output UPS-Performance and Test Requirements*; ISO/IEC: Geneva, Switzerland, 2016.
60. *Standard IEC 62040-4. Uninterruptible Power Systems (UPS)-Part 4: Environmental Aspects-Requirements and Reporting*; ISO/IEC: Geneva, Switzerland, 2013.
61. *Standard IEC62040-3. Uninterruptible Power Systems (UPS)-Part 3: Method of Specifying the Performance and Test Requirements*; ISO/IEC: Geneva, Switzerland, 2011.
62. The EMerge Alliance Data/Telecom Center Standard Creates an Integrated, Open Platform for Power, Infrastructure, Peripheral Device and Control Applications to Facilitate the Hybrid Use of AC and DC Power within Data Centers and Telecom Central Offices. Available online: https://www.emergealliance.org/standards/data-telecom/standard-faqs/ (accessed on 28 February 2021).
63. Tesla Powerwall Review. Available online: https://www.cleanenergyreviews.info/blog/tesla-powerwall-2-solar-battery-review (accessed on 25 April 2021).
64. Sonnen Documentation. Available online: http://www.sonnensupportaustralia.com.au/documentation.html (accessed on 25 April 2021).
65. Available online: https://store.enphase.com/storefront/en-us/pub/media/productattach/e/n/envoys-ds-en-us.pdf (accessed on 7 June 2021).
66. Available online: https://store.enphase.com/storefront/en-us/pub/media/productattach/e/n/enphase_encharge_3_datasheet.pdf (accessed on 7 June 2021).
67. Available online: https://store.enphase.com/storefront/en-us/pub/media/productattach/e/n/encharge_10_datasheet.pdf (accessed on 7 June 2021).
68. Victron Energy Blue Power. Available online: https://www.victronenergy.com/upload/documents/Brochure-Energy-Storage-EN_web.pdf (accessed on 25 February 2021).
69. ADARA Power Commercial Energy Storage System. Available online: http://www.adarapower.com/home/commercial-energy-storage-system/ (accessed on 25 February 2021).
70. Adara Power Introduces 20-kWh Residential Energy Storage Solution. Available online: https://www.solarpowerworldonline.com/2017/04/adara-power-introduces-20-kwh-residential-energy-storage-solution/ (accessed on 25 February 2021).
71. Schnoder Electric Hybrid Inverter/Charger XW+. Available online: https://solar.schneider-electric.com/wp-content/uploads/2020/08/DS20200812_XW-120-240-V.pdf (accessed on 25 February 2021).
72. Sunverge Energy AC-Coupled Solar Integration System (SIS). Available online: https://cdn2.hubspot.net/hubfs/2472485/WebsiteContent/Sunverge_ACSIS_NA_12092016.pdf?t=1485218396447 (accessed on 25 February 2021).
73. LG Home Battery RESU. Available online: https://www.lgessbattery.com/eu/main/main.lg (accessed on 25 April 2021).
74. Solax Triple Power Battery–LFP. Available online: https://www.solaxpower.com/triple-power-battery/ (accessed on 25 April 2021).
75. StorEdge™ On-Grid Solution. Available online: https://www.solaredge.com/solutions/self-consumption#/ (accessed on 25 April 2021).
76. 4 kWh Powervault Lead-Acid Solar Energy Storage Systems. Available online: https://www.ecopowersupplies.com/4kwh-powervault-lead-acid-solar-energy-storage-systems (accessed on 25 February 2021).
77. Purestorage, I.I. Available online: https://www.puredrive-energy.co.uk/ (accessed on 25 April 2021).
78. Duracell Energy Bank. Available online: https://www.duracellenergybank.com/ (accessed on 25 April 2021).
79. Enphase Encharge 3. Available online: https://store.enphase.com/storefront/en-us/enphase_encharge_3 (accessed on 25 April 2021).
80. Enphase Encharge 10. Available online: https://store.enphase.com/storefront/en-us/enphase_encharge_10 (accessed on 25 April 2021).
81. Available online: https://www.eaton.com/content/eaton/gb/en-gb/catalog/energy-storage/xstorage-home.html/ (accessed on 25 April 2021).
82. xStorage Home. Available online: https://www.samsungsdi.com/ess/index.html (accessed on 25 April 2021).
83. VARTA Pulse Neo. Available online: https://www.varta-ag.com/en/consumer/product-categories/energy-storage-systems/varta-pulse-neo (accessed on 25 April 2021).
84. Sunny Boy Storage 2.5. Available online: https://www.sma.de/en/products/battery-inverters/sunny-boy-storage-25.html (accessed on 25 April 2021).
85. EasySolar. Available online: https://www.victronenergy.ru/inverters-chargers/easysolar (accessed on 25 April 2021).
86. EssProTM-Battery Energy Storage. The Power to Control Energy. Available online: https://new.abb.com/docs/librariesprovider78/eventos/jjtts-2017/presentaciones-peru/(dario-cicio)-bess---battery-energy-storage-system.pdf?sfvrsn=2 (accessed on 28 February 2021).
87. Battery Energy Storage Solutions for Stable Power Supply. Available online: https://www.nidec-industrial.com/markets/renewable-energy/battery-energy-storage-solutions/ (accessed on 28 February 2021).

88. Kheraluwala, M.; Gascoigne, R.; Divan, D.; Baumann, E. Performance characterization of a high-power dual active bridge DC-to-DC converter. *IEEE Trans. Ind. Appl.* **1992**, *28*, 1294–1301. [CrossRef]
89. Stynski, S.; Luo, W.; Chub, A.; Franquelo, L.G.; Malinowski, M.; Vinnikov, D. Utility-Scale Energy Storage Systems: Converters and Control. *IEEE Ind. Electron. Mag.* **2020**, *14*, 32–52. [CrossRef]
90. Inoue, S.; Akagi, H. A Bidirectional DC–DC Converter for an Energy Storage System with Galvanic Isolation. *IEEE Trans. Power Electron.* **2007**, *22*, 2299–2306. [CrossRef]
91. Manias, S.; Ziogas, P.; Olivier, G. Bilateral DC to AC convertor using a high frequency link. *IEE Proc. B Electr. Power Appl.* **1987**, *134*, 15. [CrossRef]
92. Blinov, A.; Korkh, O.; Chub, A.; Vinnikov, D.; Peftitsis, D.; Norrga, S.; Galkin, I. High Gain DC-AC High-Frequency Link Inverter with Improved Quasi-Resonant Modulation. *IEEE Trans. Ind. Electron.* **2021**. [CrossRef]
93. Korkh, O.; Blinov, A.; Vinnikov, D.; Chub, A. Review of Isolated Matrix Inverters: Topologies, Modulation Methods and Applications. *Energies* **2020**, *13*, 2394. [CrossRef]
94. Chung, H.; Cheung, W.-L.; Tang, K. A ZCS Bidirectional Flyback DC/DC Converter. *IEEE Trans. Power Electron.* **2004**, *19*, 1426–1434. [CrossRef]
95. Lessing, M.H.; Agostini, E.; Barbi, I. An improved modulation strategy for the high-frequency-isolated DC-AC flyback converter with differential output connection. In Proceedings of the 2016 12th IEEE International Conference on Industry Applications (INDUSCON), Curitiba, Brazil, 20–23 November 2016; Institute of Electrical and Electronics Engineers (IEEE): New York, NY, USA, 2017; pp. 1–7.
96. Ponnaluri, S.; Linhofer, G.; Steinke, J.; Steimer, P.K. Comparison of single and two stage topologies for interface of BESS or fuel cell system using the ABB standard power electronics building blocks. In Proceedings of the 2005 European Conference on Power Electronics and Applications, Dresden, Germany, 11–14 September 2005; Institute of Electrical and Electronics Engineers (IEEE): New York, NY, USA, 2005.
97. Srinivasan, R.; Oruganti, R. A unity power factor converter using half-bridge boost topology. *IEEE Trans. Power Electron.* **1998**, *13*, 487–500. [CrossRef]
98. Vazquez, S.; Lukic, S.M.; Galvan, E.; Franquelo, L.G.; Carrasco, J.M. Energy Storage Systems for Transport and Grid Applications. *IEEE Trans. Ind. Electron.* **2010**, *57*, 3881–3895. [CrossRef]
99. Stanley, G.; Bradshaw, K. Precision DC-to-AC power conversion by optimization of the output current waveform-the half bridge revisited. *IEEE Trans. Power Electron.* **1999**, *14*, 372–380. [CrossRef]
100. Zhang, L.; Zhu, T.; Chen, L.; Sun, K. A systematic topology generation method for dual-buck inverters. In *2016 IEEE Energy Conversion Congress and Exposition (ECCE)*; Institute of Electrical and Electronics Engineers (IEEE): New York, NY, USA, 2017; pp. 1–6.
101. Xie, J.; Zhang, F.; Ren, R.; Wang, X.; Wang, J. A novel high power density dual-buck inverter with coupled filter inductors. In *IECON 2014-40th Annual Conference of the IEEE Industrial Electronics Society*; Institute of Electrical and Electronics Engineers (IEEE): New York, NY, USA, 2014; pp. 1111–1117.
102. Zhou, L.; Gao, F. Dual buck inverter with series connected diodes and single inductor. In *2016 IEEE Applied Power Electronics Conference and Exposition (APEC)*; Institute of Electrical and Electronics Engineers (IEEE): New York, NY, USA, 2016; pp. 2259–2263.
103. Araujo, S.V.; Zacharias, P.; Mallwitz, R. Highly Efficient Single-Phase Transformerless Inverters for Grid-Connected Photovoltaic Systems. *IEEE Trans. Ind. Electron.* **2010**, *57*, 3118–3128. [CrossRef]
104. Gu, B.; Dominic, J.; Lai, J.-S.; Chen, C.-L.; Labella, T.; Chen, B. High Reliability and Efficiency Single-Phase Transformerless Inverter for Grid-Connected Photovoltaic Systems. *IEEE Trans. Power Electron.* **2012**, *28*, 2235–2245. [CrossRef]
105. Colak, I.; Kabalci, E.; Bayindir, R. Review of multilevel voltage source inverter topologies and control schemes. *Energy Convers. Manag.* **2011**, *52*, 1114–1128. [CrossRef]
106. Liu, C.; Cai, X.; Chen, Q. Self-Adaptation Control of Second-Life Battery Energy Storage System Based on Cascaded H-Bridge Converter. *IEEE J. Emerg. Sel. Top. Power Electron.* **2018**, *8*, 1428–1441. [CrossRef]
107. Ling, Z.; Zhang, Z.; Li, Z.; Li, Y. State-of-charge balancing control of battery energy storage system based on cascaded H-bridge multilevel inverter. In Proceedings of the 2016 IEEE 8th International Power Electronics and Motion Control Conference (IPEMC-ECCE Asia); Institute of Electrical and Electronics Engineers (IEEE): New York, NY, USA, 2016; pp. 2310–2314.
108. Yoscovich, I.; Glovinsky, T.; Sella, G.; Galin, Y. SolarEdge Patent for HD-Wave Inverters-Distributed Power System Using Direct Current Power Sources. U.S. Patent 9368964B2, 14 June 2016.
109. Demetriades, G.D. Modular Multilevel Converter with Cell-Connected Battery Storages. European Patent EP2695273B1, 25 November 2015.
110. Moradpour, M.; Ghani, P.; Pirino, P.; Gatto, G. A GaN-Based Battery Energy Storage System for Three-Phase Residential Application with Series-Stacked Devices and Three-Level Neutral Point Clamped Topology. In *2019 1st International Conference on Energy Transition in the Mediterranean Area (SyNERGY MED)*; Institute of Electrical and Electronics Engineers (IEEE): New York, NY, USA, 2019; pp. 1–6.
111. Abronzini, U.; Attaianese, C.; Di Monaco, M.; Tomasso, G.; Damiano, A.; Porru, M.; Serpi, A. A Dual-Source DHB-NPC Power Converter for Grid Connected Split Battery Energy Storage System. In *2018 IEEE Energy Conversion Congress and Exposition (ECCE)*; Institute of Electrical and Electronics Engineers (IEEE): New York, NY, USA, 2018; pp. 2483–2488.

112. Trintis, I.; Teodorescu, R.; Munk-Nielsen, S. Single stage grid converters for battery energy storage. In Proceedings of the 5th IET International Conference on Power Electronics, Machines and Drives (PEMD 2010), Brighton, UK, 19–21 April 2010; Institution of Engineering and Technology (IET): London, UK, 2010; p. 234.
113. Bragard, M.; Soltau, N.; Thomas, S.; De Doncker, R.W. The Balance of Renewable Sources and User Demands in Grids: Power Electronics for Modular Battery Energy Storage Systems. *IEEE Trans. Power Electron.* **2010**, *25*, 3049–3056. [CrossRef]
114. Cheng, Y.; Qian, C.; Crow, M.L.; Pekarek, S.; Atcitty, S. A Comparison of Diode-Clamped and Cascaded Multilevel Converters for a STATCOM with Energy Storage. *IEEE Trans. Ind. Electron.* **2006**, *53*, 1512–1521. [CrossRef]
115. Barth, C.B.; Assem, P.; Foulkes, T.; Chung, W.H.; Modeer, T.; Lei, Y.; Pilawa-Podgurski, R.C.N. Design and Control of a GaN-Based, 13-Level, Flying Capacitor Multilevel Inverter. *IEEE J. Emerg. Sel. Top. Power Electron.* **2019**, *8*, 2179–2191. [CrossRef]
116. Modeer, T.; Pallo, N.; Foulkes, T.; Barth, C.B.; Pilawa-Podgurski, R.C.N. Design of a GaN-Based Interleaved Nine-Level Flying Capacitor Multilevel Inverter for Electric Aircraft Applications. *IEEE Trans. Power Electron.* **2020**, *35*, 12153–12165. [CrossRef]
117. Siwakoti, Y.P.; Peng, F.Z.; Blaabjerg, F.; Loh, P.C.; Town, G.E. Impedance-Source Networks for Electric Power Conversion Part I: A Topological Review. *IEEE Trans. Power Electron.* **2015**, *30*, 699–716. [CrossRef]
118. Mande, D.; Trovão, J.P.; Ta, M.C. Comprehensive Review on Main Topologies of Impedance Source Inverter Used in Electric Vehicle Applications. *World Electr. Veh. J.* **2020**, *11*, 37. [CrossRef]
119. You, K.; Rahman, M.F. A Matrix–Z-Source Converter with AC–DC Bidirectional Power Flow for an Integrated Starter Alternator System. *IEEE Trans. Ind. Appl.* **2009**, *45*, 239–248. [CrossRef]
120. Matiushkin, O.; Husev, O.; Strzelecki, R.; Ivanets, S.; Fesenko, A. Novel single-stage buck-boost inverter with unfolding circuit. In Proceedings of the 2017 IEEE First Ukraine Conference on Electrical and Computer Engineering (UKRCON), Kyiv, Ukraine, 29 May–2 June 2017; Institute of Electrical and Electronics Engineers (IEEE): New York, NY, USA, 2017; pp. 538–543.
121. Han, B.; Lai, J.-S.; Kim, M. Bridgeless Cuk-Derived Single Power Conversion Inverter with Reactive-Power Capability. *IEEE Trans. Power Electron.* **2019**, *35*, 2629–2645. [CrossRef]
122. Pal, A.; Basu, K. A Single-Stage Soft-Switched Isolated Three-Phase DC–AC Converter with Three-Phase Unfolder. *IEEE Trans. Power Electron.* **2019**, *35*, 3601–3615. [CrossRef]
123. Tytelmaier, K.; Husev, O.; Veligorskyi, O.; Yershov, R. A review of non-isolated bidirectional dc-dc converters for energy storage systems. In *2016 II International Young Scientists Forum on Applied Physics and Engineering (YSF)*; Institute of Electrical and Electronics Engineers (IEEE): New York, NY, USA, 2016; pp. 22–28.
124. Anzola, J.; Aizpuru, I.; Romero, A.A.; Loiti, A.A.; Lopez-Erauskin, R.; Sevil, J.S.A.; Bernal, C. Review of Architectures Based on Partial Power Processing for DC-DC Applications. *IEEE Access* **2020**, *8*, 103405–103418. [CrossRef]
125. Pape, M.; Kazerani, M. An Offshore Wind Farm with DC Collection System Featuring Differential Power Processing. *IEEE Trans. Energy Convers.* **2019**, *35*, 222–236. [CrossRef]
126. Pape, M.; Kazerani, M. Turbine Startup and Shutdown in Wind Farms Featuring Partial Power Processing Converters. *IEEE Open Access J. Power Energy* **2020**, *7*, 254–264. [CrossRef]
127. Iyer, V.M.; Gulur, S.; Gohil, G.; Bhattacharya, S. An Approach towards Extreme Fast Charging Station Power Delivery for Electric Vehicles with Partial Power Processing. *IEEE Trans. Ind. Electron.* **2020**, *67*, 8076–8087. [CrossRef]
128. Schaef, C.; Stauth, J.T. Multilevel Power Point Tracking for Partial Power Processing Photovoltaic Converters. *IEEE J. Emerg. Sel. Top. Power Electron.* **2014**, *2*, 859–869. [CrossRef]
129. Shenoy, P.S. Improving Performance, Efficiency, and Reliability of DC/DC Conversion Systems by Differential Power Processing. Ph.D. Thesis, Graduate College of the University of Illinois at Urbana-Champaign, Champaign, IL, USA, 2012.
130. Zientarski, J.R.R.; Martins, M.L.D.S.; Pinheiro, J.R.; Hey, H.L. Evaluation of Power Processing in Series-Connected Partial-Power Converters. *IEEE J. Emerg. Sel. Top. Power Electron.* **2019**, *7*, 343–352. [CrossRef]
131. Zientarski, J.R.R.; Martins, M.L.D.S.; Pinheiro, J.R.; Hey, H.L. Series-Connected Partial-Power Converters Applied to PV Systems: A Design Approach Based on Step-Up/Down Voltage Regulation Range. *IEEE Trans. Power Electron.* **2018**, *33*, 7622–7633. [CrossRef]
132. Zapata, J.W.; Kouro, S.; Carrasco, G.; Renaudineau, H.; Meynard, T.A. Analysis of Partial Power DC–DC Converters for Two-Stage Photovoltaic Systems. *IEEE J. Emerg. Sel. Top. Power Electron.* **2019**, *7*, 591–603. [CrossRef]
133. Agamy, M.S.; Harfman-Todorovic, M.; Elasser, A.; Chi, S.; Steigerwald, R.L.; Sabate, J.A.; McCann, A.J.; Zhang, L.; Mueller, F.J. An Efficient Partial Power Processing DC/DC Converter for Distributed PV Architectures. *IEEE Trans. Power Electron.* **2014**, *29*, 674–686. [CrossRef]
134. Marti-Arbona, E.; Mandal, D.; Bakkaloglu, B.; Kiaei, S. PV panel power optimization using sub-panel MPPT. In Proceedings of the 2015 IEEE Applied Power Electronics Conference and Exposition (APEC), Charlotte, NC, USA, 15–19 March 2015; Institute of Electrical and Electronics Engineers (IEEE): New York, NY, USA, 2015; pp. 235–238.
135. Loera-Palomo, R.; Morales-Saldaña, J.A.; Palacios-Hernández, E. Quadratic step-down dc–dc converters based on reduced redundant power processing approach. *IET Power Electron.* **2013**, *6*, 136–145. [CrossRef]
136. Spiazzi, G. Reduced redundant power processing concept: A reexamination. In Proceedings of the 2016 IEEE 17th Workshop on Control and Modeling for Power Electronics (COMPEL), Trondheim, Norway, 27–30 June 2016; pp. 1–8. [CrossRef]
137. Xue, F.; Yu, R.; Huang, A. Fractional converter for high efficiency high power battery energy storage system. In Proceedings of the 2017 IEEE Energy Conversion Congress and Exposition (ECCE), Cincinnati, OH, USA, 1–5 October 2017; Institute of Electrical and Electronics Engineers (IEEE): New York, NY, USA, 2017; pp. 5144–5150.

138. Xue, F.; Yu, R.; Huang, A. A Family of Ultrahigh Efficiency Fractional dc–dc Topologies for High Power Energy Storage Device. *IEEE J. Emerg. Sel. Top. Power Electron.* **2021**, *9*, 1420–1427. [CrossRef]
139. Neumayr, D.; Knabben, G.C.; Varescon, E.; Bortis, D.; Kolar, J.W. Comparative Evaluation of a Full- and Partial-Power Processing Active Power Buffer for Ultracompact Single-Phase DC/AC Converter Systems. *IEEE J. Emerg. Sel. Top. Power Electron.* **2021**, *9*, 1994–2013. [CrossRef]
140. Bala, S.; Tengner, T.; Rosenfeld, P.; Delince, F. The effect of low frequency current ripple on the performance of a Lithium Iron Phosphate (LFP) battery energy storage system. In *2012 IEEE Energy Conversion Congress and Exposition (ECCE)*; Institute of Electrical and Electronics Engineers (IEEE): New York, NY, USA, 2012; pp. 3485–3492.
141. Brand, M.J.; Hofmann, M.H.; Schuster, S.S.; Keil, P.; Jossen, A. The Influence of Current Ripples on the Lifetime of Lithium-Ion Batteries. *IEEE Trans. Veh. Technol.* **2018**, *67*, 10438–10445. [CrossRef]
142. Husev, O.; Matiushkin, O.; Roncero-Clemente, C.; Blaabjerg, F.; Vinnikov, D. Novel Family of Single-Stage Buck–Boost Inverters Based on Unfolding Circuit. *IEEE Trans. Power Electron.* **2018**, *34*, 7662–7676. [CrossRef]
143. Zhao, B.; Abramovitz, A.; Liu, C.; Yang, Y.; Huangfu, Y. A Family of Single-Stage, Buck-Boost Inverters for Photovoltaic Applications. *Energies* **2020**, *13*, 1675. [CrossRef]
144. Oguchi, K.; Ikawa, E.; Tsukiori, Y. A three-phase sine wave inverter system using multiple phase-shifted single-phase resonant inverters. *IEEE Trans. Ind. Appl.* **1993**, *29*, 1076–1083. [CrossRef]
145. Rabkowski, J.; Blinov, A.; Zinchenko, D.; Wrona, G.; Zdanowski, M. Grid-frequency Vienna rectifier and isolated current-source DC-DC converters for efficient off-board charging of electric vehicles. In Proceedings of the 2020 22nd European Conference on Power Electronics and Applications (EPE'20 ECCE Europe), Piscataway, NJ, USA, 7–11 September 2020; IEEE: New York, NY, USA, 2020.
146. Jacobson, B.S.; Holmansky, E.N. Methods and Apparatus for Three-Phase Inverter with Reduced Energy Storage. U.S. Patent 7839023B2, 23 November 2010.
147. Chen, W.W.; Zane, R.; Corradini, L. Isolated Bidirectional Grid-Tied Three-Phase AC–DC Power Conversion Using Series-Resonant Converter Modules and a Three-Phase Unfolder. *IEEE Trans. Power Electron.* **2017**, *32*, 9001–9012. [CrossRef]
148. Van Soest, H. Peer-to-peer electricity trading: A review of the legal context. *Compet. Regul. Netw. Ind.* **2018**, *19*, 180–199. [CrossRef]
149. Zhang, D.; Guacci, M.; Kolar, J.W.; Everts, J. Synergetic Control of a 3-Φ Buck-Boost Current DC-Link EV Charger Considering Wide Output Range and Irregular Mains Conditions. In Proceedings of the 2020 IEEE 9th International Power Electronics and Motion Control Conference (IPEMC2020-ECCE Asia), Nanjing, China, 31 May–3 June 2020; pp. 1688–1695.
150. Husev, O.; Roncero-Clemente, C.; Romero-Cadaval, E.; Vinnikov, D.; Stepenko, S. Single phase three-level neutral-point-clamped quasi-Z-source inverter. *IET Power Electron.* **2015**, *8*, 1–10. [CrossRef]
151. Panfilov, D.; Husev, O.; Blaabjerg, F.; Zakis, J.; Khandakji, K. Comparison of three-phase three-level voltage source inverter with intermediate dc–dc boost converter and quasi-Z-source inverter. *IET Power Electron.* **2016**, *9*, 1238–1248. [CrossRef]
152. Baier, T.; Piepenbreier, B. Bidirectional magnetically coupled T-Source Inverter for extra low voltage application. In Proceedings of the 2016 IEEE Applied Power Electronics Conference and Exposition (APEC), Long Beach, CA, USA, 20–24 March 2016; pp. 2897–2904.
153. Hu, S.; Liang, Z.; He, X. Ultracapacitor-Battery Hybrid Energy Storage System Based on the Asymmetric Bidirectional Z -Source Topology for EV. *IEEE Trans. Power Electron.* **2016**, *31*, 7489–7498. [CrossRef]

 energies

Article

Design of an Effective State of Charge Estimation Method for a Lithium-Ion Battery Pack Using Extended Kalman Filter and Artificial Neural Network

Van Quan Dao [1], Minh-Chau Dinh [2], Chang Soon Kim [2], Minwon Park [1], Chil-Hoon Doh [3], Jeong Hyo Bae [3], Myung-Kwan Lee [4], Jianyong Liu [5] and Zhiguo Bai [5,*]

1 Department of Electrical Engineering, Changwon National University, Changwon 51140, Korea; quandao.hust98@gmail.com (V.Q.D.); paku@changwon.ac.kr (M.P.)
2 Institute of Mechatronics, Changwon National University, Changwon 51140, Korea; thanchau7787@gmail.com (M.-C.D.); ee.cskim@gmail.com (C.S.K.)
3 Distributed Power System Research Center, Korea Electrotechnology Research Institute, Changwon 51543, Korea; chdoh@keri.re.kr (C.-H.D.); jhbae@keri.re.kr (J.H.B.)
4 Battery Solution Co., Ltd., Jeonnam 58324, Korea; kem4328@naver.com
5 IES Co., Ltd., Busan 46744, Korea; newsir86@gmail.com
* Correspondence: kawabai@gmail.com; Tel.: +82-10-8310-2595

Citation: Dao, V.Q.; Dinh, M.-C.; Kim, C.S.; Park, M.; Doh, C.-H.; Bae, J.H.; Lee, M.-K.; Liu, J.; Bai, Z. Design of an Effective State of Charge Estimation Method for a Lithium-Ion Battery Pack Using Extended Kalman Filter and Artificial Neural Network. *Energies* **2021**, *14*, 2634. https://doi.org/10.3390/en14092634

Academic Editor: Andrei Blinov

Received: 6 April 2021
Accepted: 27 April 2021
Published: 4 May 2021

Abstract: Currently, Lithium-ion batteries (LiB) are widely applied in energy storage devices in smart grids and electric vehicles. The state of charge (SOC) is an indication of the available battery capacity, and is one of the most important factors that should be monitored to optimize LiB's performance and improve its lifetime. However, because the SOC relies on many nonlinear factors, it is difficult to estimate accurately. This paper presented the design of an effective SOC estimation method for a LiB pack Battery Management System (BMS) based on Kalman Filter (KF) and Artificial Neural Network (ANN). First, considering the configuration and specifications of the BMS and LiB pack, an ANN was constructed for the SOC estimation, and then the ANN was trained and tested using the Google TensorFlow open-source library. An SOC estimation model based on the extended KF (EKF) and a Thevenin battery model was developed. Then, we proposed a combined mode EKF-ANN that integrates the estimation of the EKF into the ANN. Both methods were evaluated through experiments conducted on a real LiB pack. As a result, the ANN and KF methods showed maximum errors of 2.6% and 2.8%, but the EKF-ANN method showed better performance with less than 1% error.

Keywords: Artificial neural network; battery management system; Kalman filter; lithium-ion battery; state of charge estimation

1. Introduction

Energy storage systems are emerging as the biggest concern for modern smart grids and electric vehicles (EV), and the lithium-ion battery (LiB) technology is an efficient solution for energy storage applications with the advantages of long cycle life, large capacity and no memory effect. Already commercialized and matured for consumer electronic applications, the LiB is being positioning itself as a leading technology platform for plug-in hybrid electric vehicles (PHEVs) and all EVs [1,2]. It is also widely used in large facilities to support energy storage [3], load-leveling and peak shaving in the power grid [4], frequency regulation [5], and to reduce network load and capacity payments [6] in the smart grids.

In order for the LiBs to work as expected, a battery management system (BMS) must be designed for tracking and controlling the current level of battery energy. The BMS is defined as an electronic equipment that manages a rechargeable battery (single cell or battery pack). The main functions of the BMS are to monitor, compute, communicate, protect, and optimize [7]. At this point, the estimate of state of charge (SOC) is one of the critical functions in the BMS. The SOC is defined as the percentage of the available capacity to the rated capacity of the battery, and many issues with the LiB, such as capacity degradation, increased maintenance costs, rapid aging, serious equipment failures, and even dangerous accidents, are related to incorrect SOC estimates [8]. Therefore, an accurate estimation of the SOC is very important for optimizing battery performance, including extending battery life and preventing permanent damage to the batteries.

In general, the battery SOC nonlinearly depends on several factors including current, voltage, temperature, and battery aging [9]. Therefore, an accurate estimate of the SOC is quite complicated. Various techniques have been presented to estimate the SOC of a battery cell or a battery pack. Key technologies include discharge tests, open-circuit voltage measurement, Coulomb counting, inherent resistance measurement, and intelligent SOC estimation methods [10,11]. Intelligent computation techniques such as artificial neural network (ANN) and Kalman filter (KF) have been developed for EV applications [12–16]. Compared to other techniques, they have several advantages such as high accuracy, real-time calculation, simple current and voltage measurements. Specifically, these techniques are highly adaptable to the dynamic behaviors of batteries due to their self-learning ability. However, there are still some issues that need to be studied. Applying KF requires accurate battery modeling, and important factors such as temperature and SOC that may affect the internal parameters of the battery model are not yet considered. Using an ANN requires a large amount of training data that can lead to a large dimension and high computation of the network when implemented in a BMS. Therefore, it is necessary to design a practical BMS to properly analyze and evaluate the operational characteristics of the SOC estimation methods.

In this paper, an effective SOC estimation method was designed and implemented in a smart BMS for a LiB pack based on the extended KF (EKF) and ANN. First, the structure and specifications of the smart BMS and LiB pack were summarized, and the design process of the ANN was described in detail. The ANN was then trained and tested for SOC estimation using real battery data sets. Next, we developed an SOC estimation algorithm based on the EKF and a Thevenin battery model. Finally, we proposed a combination model of EKF and ANN (EKF-ANN) to compensate for the shortcomings of the above two methods. To evaluate the effectiveness of the SOC estimation method, the proposed methods were experimentally verified and compared with each other. As a result, the proposed ANN and EKF methods showed an error of 2.6% and 2.8%, respectively, and the SOC estimation error when using the EKF-ANN was significantly improved to less than 1%. The results show that the proposed SOC estimation method satisfies the requirements of the BMS for LiB packs.

2. Review of SOC Estimation Methods

An accurate estimate of the SOC plays an important role in a credible BMS, but the SOC cannot be measured directly. The SOC is associated with direct measurements such as current, voltage, temperature, and it can be extracted based on intrinsic relations or control theory of the battery. Many techniques have been proposed to estimate the battery SOC. In this section, we discuss some popular SOC estimation methods and compare them with each other.

2.1. Open Circuit Voltage Method

This method calculates the SOC or the remaining capacity of the battery based on the measured open-circuit voltage (OCV). Each battery has a corresponding curve between the SOC and OCV, from which one can determine the other. However, in order to get a stable voltage, the battery must be rested for a long time under no load. Moreover, since the OCV-SOC curve is sensitive to various temperatures and discharge rates, the method is only effective in estimating the SOC at the early and end stages of the charging and discharging process after the battery has been disconnected from the load for a long time [17,18].

2.2. Coulomb Counting Method

Coulomb counting, which is also called Ampare-hour (Ah) counting, is the most common technique for estimating the SOC based on the integration on time of the charge and discharge current values. However, the initial SOC is difficult to determine at the starting state. Even though we can gain the SOC from the record of the BMS or OCV look-up table, but the accuracy is hard to ensure [19]. Additionally, the SOC calculation is based only on the measurement of current without considering the measurement noise. Over time, errors will be accumulated due to the integration factor, and this is the reason why this method is prone to errors.

2.3. Impedance-Based Method

There is a dependency between the SOC and the impedance of a battery, and thus the SOC can be considered a function of battery impedance change [20]. However, the impedance varies significantly with the aging status of a battery; thus, this technology is no longer a good indicator for the SOC. Furthermore, the sensitivity of the battery impedance on the temperature is very high; thereby a high accuracy of SOC estimation is impossible to maintain for batteries in EVs due to quick temperature change during the driving process.

2.4. Kalman Filter-Based Method

The KF is a method for determining the internal states of any dynamic process, in which the mean of the squared error is minimized. Its target is to obtain accurate information from inaccurate data. This method can be utilized to calculate the SOC in real-time by using the terminal current and voltage measurements [21–23]. It is suitable for the SOC estimation of EVs in which the battery current is unstable [24]. However, it has high demands for the battery modelling and computational capability [25,26].

2.5. Artificial Neural Network-Based Method

The ANN is an intelligent technology, which has a strong self-learning and high adaptability, and this technique is very useful for researching complex nonlinear system models. For the SOC estimation, the ANN is able to be applied in all battery systems without the information of cell internal structure, as long as the battery dataset for training the network is available [27,28]. Also, the ANN has the ability to estimate the SOC without the initial SOC.

2.6. Fuzzy Logic-Based Method

This method is based on simulating the fuzzy thinking of a human being using the fuzzy logic based on a large number of test curves, experience and reliable fuzzy logic theories, and finally to perform the SOC estimation [29]. It requires a complete understanding of the battery itself and relatively large computations. However, the battery parameters significantly vary with the battery lifetime, and so the SOC estimation may not be accurate enough. It is only suitable for static battery characteristics and practically inapplicable for LiBs in the EVs [30].

In this paper, we decided to develop an effective SOC estimation method based on the ANN and KF.

3. Design of SOC Estimation Methods for a Smart BMS

3.1. Configuration of the Smart BMS and LiB Pack

A BMS is a device including hardware and software which controls the operating conditions of the battery to extend battery life, ensure safety, and accurately predict the various states of the battery. To ensure this, the BMS has several functions to control and monitor the battery states at various battery cells, battery modules, and battery pack levels.

The estimation of the SOC is a core function in a BMS, but it is still a challenge to accurately estimate in online-real time, as battery characteristics change with the degree of aging and strong nonlinear and complex electrochemical reactions of the battery. The SOC can be estimated by continuously measuring terminal voltage, current and temperature of the battery. For a battery system in EVs and smart grids, the current trend is the design of a smart BMS, which includes researches in the field of artificial intelligence (AI) utilized for the battery SOC estimation. In large battery pack applications, the predictability and adaptability of BMS are especially important. In this study, in order to verify the performance of the SOC estimation methods through experiments, we developed a smart BMS module based on the master and slave topology design for the LiB pack as shown in Figure 1.

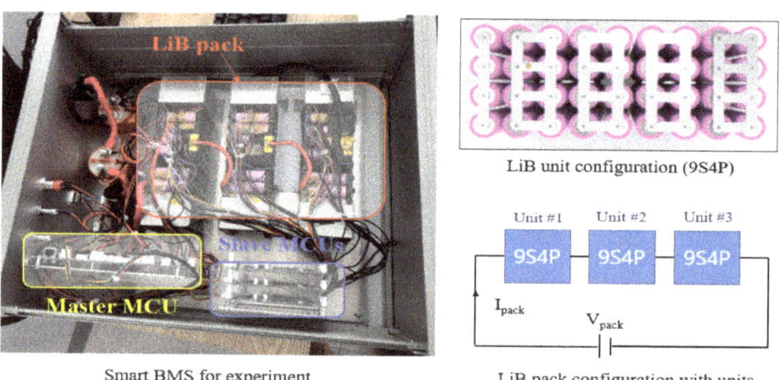

Figure 1. Configurations of the smart BMS and LiB pack.

The LiB pack consisted of 3 module units connected in series, and each module unit consisted of 36 cells with a 9S4P configuration, meaning 9 cells in series and 4 cells in parallel. There are 3 slave microcontroller unit (MCU) boards to manage the module units. The main MCU board, which is the master, coordinates two-way communication between the master and the slave and executes battery management actions. The detailed specifications of the smart BMS and LiB pack are described in Table 1. We will develop SOC estimation methods with an error target of 3% as shown in Table 2.

Table 1. Detailed specifications of the smart BMS and LiB pack.

Items	Specifications	Contents
1. BMS hardware	Master BMS MCU	32-bit STM32F207
	Slave BMS MCU	32-bit STM32F103
	External communication	UART
	Master-Slave communication	CAN 2.0
	Power supply	12 V
	Temperature sensors	4 ea.
	Firmware tool	ST-Link Download
2. LiB cell	Model	18,650 35E
	Capacity	3.4 Ah (1 C)
	Cell voltage	2.5~4.2 V

Table 1. *Cont.*

Items	Specifications	Contents
3. LiB unit module	Configuration	9-series and 4-parallel (9S4P)
	Voltage/Capacity	32.4 V/10.2 Ah
	Energy	330.48 Wh
4. LiB pack	Configuration	3 LiB units in series
	Pack voltage	67.5~113.4 V
	Capacity	10.2 Ah
	Energy	991.44 Wh

Table 2. Design targets of the smart BMS.

Items	Standard [31,32]	Target	Unit	Evaluation Method
Active Cell Balancing	±10	±8	mV	Digital multimeter
Voltage measurement accuracy	±5	±1	mV	Digital multimeter
Current measurement accuracy	±2	±1	%	Charger/Discharger
Temperature measurement accuracy	±1	±0.5	°C	Thermostat
SOC estimation error	±5	±3	%	Charger/Discharger

3.2. *SOC Estimation Method Using an ANN*

3.2.1. Design Process for the ANN

An ANN is a computational model constructed by a set of individual processing units, which are called the artificial neurons. These neurons are interconnected by weights. The ANN is a universal approximator that is able to model any nonlinear function with the desired precision [33]. The network is configured in layers, with the input layer receiving inputs and the output layer generating outputs. The middle layers, called the hidden layers, have no connection with the outside. The neurons are responsible for connecting each layer, thereby they are the central component of the ANN. The basic configuration of a neuron in the network is described in Figure 2. The vector of the input signals is denoted by $X = [x_1, x_2, x_3, \ldots, x_n]$, $n \in N$, the neuron weights by $W = [W_1, W_2, W_3, \ldots, W_n]$, *net* is the multiplication response of weights with the input signals, b_1 is an external parameter called as the bias, f is the activation function, and y_m is the output signal of the neuron.

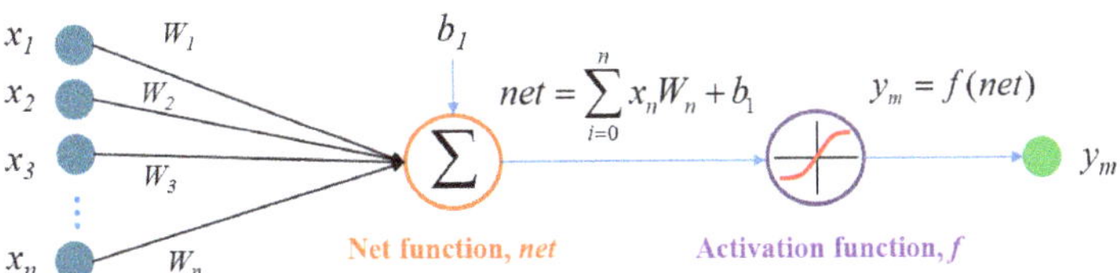

Figure 2. Configuration of a neuron in the ANN.

Table 3 shows the design process and design factors of an ANN for the SOC estimation. It starts from collecting and pre-processing the data, then designing the neural network in terms of network topology, configuration, activation function, loss function and metrics, loss optimizer, and learning rate. Finally, the training and validation of the network are implemented.

Table 3. Design process and design factors for the ANN.

Step	Contents	Remarks
1	Data collection and pre-processing for training the network	- Training dataset (%) - Validation dataset (%) - Test dataset (%)
2	Selecting the network topology and configuring the layers and neurons	- Input layer: set the inputs - Number of hidden layers, and number of neurons in each layer - Output layer: set the output
3	Selecting the learning rate, α	- $0 < \alpha < 1$
4	Selecting the activation function	- Sigmoid, Linear, Tanh, ReLu, …
5	Selecting the loss function and metrics for the training process	- Mean squared error (MSE) - Mean absolute error (MAE) - Metrics = [MSE, MAE]
6	Selecting the learning function (optimizer)	- Adam, AdaDelta, SGD, …

3.2.2. Design of the ANN for the SOC Estimation

The historical battery dataset was collected from the designed LiB pack by the company Battery Solution Co., Ltd. in South Korea. The dataset consists of the test time in seconds, the pack current in ampere, the pack terminal voltage in volts, the pack temperature, and the cumulative charge/discharge capacity of the battery pack in ampere-hour. The LiB pack was tested under 0–100% of the SOC condition for 20 cycles at room temperature. The data was collected with a sampling time of 1 s. The battery pack was charged and discharged with a constant current of 1.7 A (0.5 C rate). The battery charge and discharge processes correspond to the plus (+) and minus (−) signs of the current. The temperature was measured at the center of the battery pack by a thermal sensor with an accuracy of ±0.5 °C, and the temperature range was from 19 °C to 42 °C. From the battery pack dataset, the average values of voltage and current in one cell were calculated. The cell voltage range was from 2.6 V to 4.2 V. The actual SOC was also calculated offline by the integration on time of current to use for training and testing the ANN model. The dataset must be processed to obtain a satisfactory ANN, all trash data was deleted. The ANN was trained to estimate the SOC of one cell.

The next step was to design the network topology for the ANN, and the Multi-Layer Perception (MLP) network was selected. The inputs of the network were the current, voltage, and temperature, and the output was the battery SOC. Figure 3 describes the relationships between the inputs and outputs of the designed ANN. The SOC has strong and nonlinear relationships with the voltage and temperature, which can be expressed as exponential functions. Thus, the nonlinear activation functions including hyperbolic tangent (tanh) and logistic (sigmoid) were considered for the ANN. The learning rate, α, is an amount by which the weights are updated during the training process. In the proposed network, the learning rate was set at 0.001.

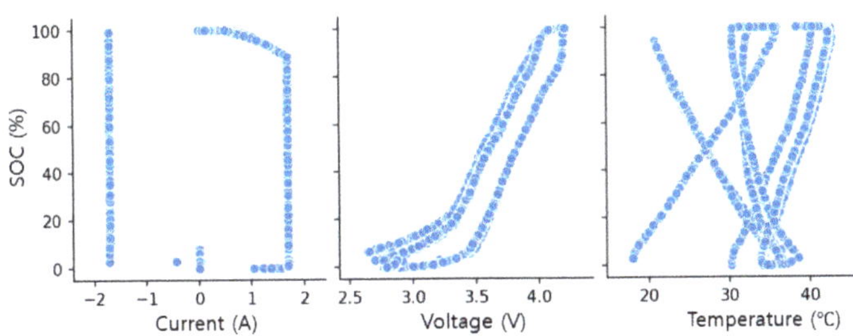

Figure 3. Relationships between the inputs and output of the ANN.

After tuning the parameters and testing several configurations of the network, the most suitable specifications for the ANN are described in Table 4.

Table 4. Final specifications of the ANN for the SOC estimation.

Specifications	ANN
Number of inputs of the network	3
Number of hidden layers of the network	4
Number of outputs of the network	1
Number of neurons in each hidden layer	64
Activation function in hidden layers	Sigmoid
Activation function in the output layer	Softplus
Adapting learning function	Adam
Learning rate	0.001
Loss function	MSE
Accuracy metrics	[MSE, MAE]

The battery dataset contained approximately 250,000 samples, corresponding to the data measured at each sampling time. Each sample consisted of three input variables (battery voltage, current and temperature) and the actual SOC. Before training the network, the dataset was normalized and randomized to separate them into three sets: training, validation, and testing. The data in each set were different from each other and split evenly into three input variables. The amount of sample data for the training, validation, and testing were 80%, 10%, and 10% of the total samples, respectively. In this study, the ANN was designed, trained, and tested by using the Keras framework in the TensorFlow open-source library of Google. The ANN was trained during 500 epochs. As a result, the training performance achieved an MSE of approximately 0.5 and an MAE of approximately 0.43 in the final epoch as shown in Figure 4. From the epoch of 150, the training losses almost match the validating losses.

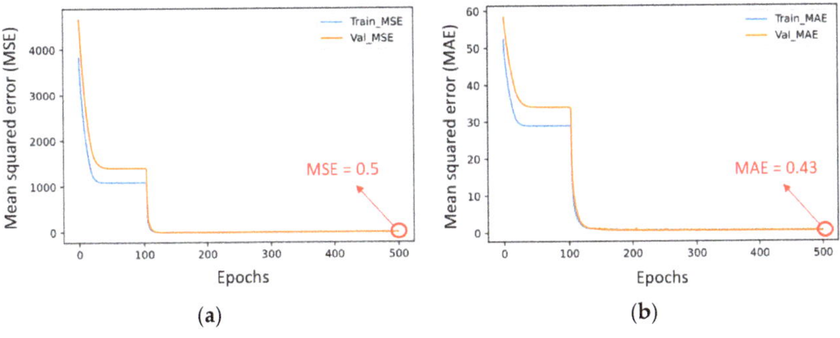

Figure 4. Training performances of the ANN: (**a**) Mean squared error (MSE); and (**b**) Mean absolute error (MAE).

After the training process, the model testing was performed to compare the estimated results and the actual results as described in Figure 5. The trained ANN was tested randomly with 5,000 samples, and the maximum and average errors of the estimated SOC were 2.3% and 0.34%, respectively. This result is satisfied with the target of SOC estimation.

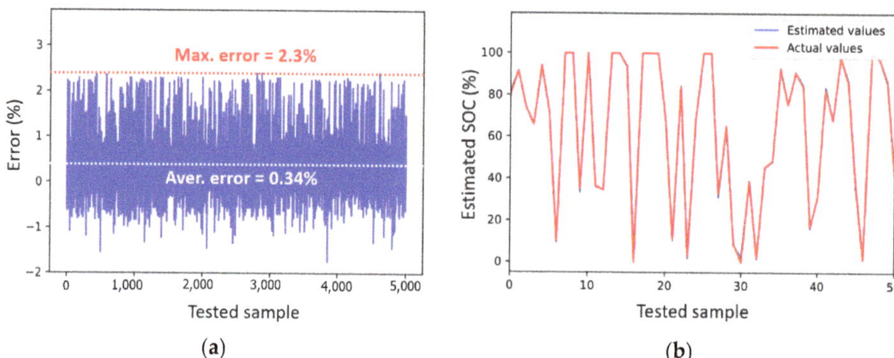

Figure 5. Test results of the trained ANN with 5,000 random samples: (**a**) error of the estimated SOC, and (**b**) estimated and actual SOCs.

When the training process was completed, the weights and bias of the network were saved and exported as matrices to implement the SOC estimation function in the BMS for experiments.

3.3. SOC Estimation Method Using the EKF

In 1960, Kalman successfully introduced the state space into filter theory and first proposed the KF. Basically, the standard KF is only appropriate for linear dynamic systems to obtain the best and unbiased estimate of the state variables. However, the battery is a non-linear system; thus, it needs to be linearized and approximated to a linear system. At this point, the system states can be estimated by the KF. This method is called the extended Kalman Filter (EKF) [34]. To implement the EKF, a battery model is first developed to define internal state variables, state-space equations to build the mathematical model. Moreover, based on the external variables, such as terminal voltage, current, and temperature, the battery model is able to determine the internal state variables such as the internal resistance, electromotive force, capacitance and SOC.

3.3.1. Battery Model for the LiB

In this study, the Thevenin model was chosen as a LiB model. This model type has good performance and can accurately simulate the dynamic characteristics of the LiB [35]. Figure 6 describes the equivalent circuit of the Thevenin battery model, which consists of an ideal source of voltage V_{OC} (OCV), an internal resistance R_0, a polarization resistance R_p, and a capacitor C_p. The terminal voltage and current of the battery are denoted by V_L and I_L, respectively. The voltage and current flowing through the branch R_pC_p are denoted by V_p and I_p, respectively.

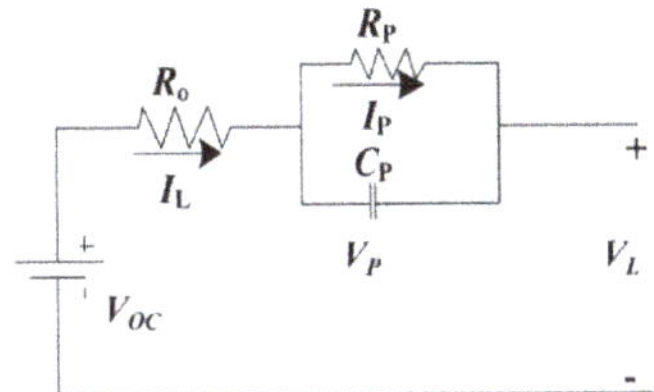

Figure 6. Equivalent circuit of the Thevenin model for the LiB.

Based on the Kirchhoff's laws, the capacitor voltage variation, its current relationship and the characteristics of the Thevenin model can be expressed as follows:

$$\dot{V_P} = \frac{I_L}{C_P} - \frac{V_P}{R_P C_P} \tag{1}$$

$$V_L = V_{OC} - V_P - I_L R_O \tag{2}$$

From Equations (1) and (2), where $\tau = R_p C_p$ is the time constant and Δt is the sampling time, a discrete-time system for the V_L and V_p is built as shown in the Equations (3) and (4).

$$V_P(k) = \exp(\frac{-\Delta t}{\tau}) \times V_P(k-1) + (1 - \exp(\frac{-\Delta t}{\tau})) \times I_L(k-1) R_P(k) \tag{3}$$

$$V_L(k) = V_{OC}(k) - V_P(k) - I_L(k) R_O(k) \tag{4}$$

The V_{OC}, R_O, R_p and C_p are nonlinear functions of the battery SOC [36], thus the Equations (3) and (4) can also be expressed as below.

$$V_P(k) = \exp(\frac{-\Delta t}{\tau}) \times V_P(k-1) + (1 - \exp(\frac{-\Delta t}{\tau})) \times I_L(k-1) R_P(SOC(k)) \tag{5}$$

$$V_L(k) = V_{OC}(SOC(k)) - V_P(k) - I_L(k) R_O(SOC(k)) \tag{6}$$

The SOC describes the relationship between the remaining and the maximum capacity available in the battery and can be described as Equation (7):

$$SOC(k) = SOC(k-1) - \frac{I_L(k) \times \Delta t}{C_a} \tag{7}$$

where C_a is the nominal capacity of the battery, and the $SOC(k-1)$ is the priority SOC. By using the Equations (5)–(7), we can express the state equation for the nonlinear battery system as follows:

$$\begin{cases} x_k = A_{k-1} x_{k-1} + B_{k-1} u_{k-1} + w_{k-1} \\ y_k = C_k x_k + D_k u_k + v_k \end{cases} \tag{8}$$

with, $x_k = \begin{bmatrix} V_P(k) \\ SOC(k) \end{bmatrix}$, $A_k = \begin{bmatrix} 1 & 0 \\ 0 & \exp\left(\frac{-\Delta t}{\tau}\right) \end{bmatrix}$, $B_k = \begin{bmatrix} R_P \exp\left(\frac{-\Delta t}{\tau}\right) \\ \frac{\Delta t}{C_a} \end{bmatrix}$, $C_k = \begin{bmatrix} \frac{dV_{oc}}{dSOC} - \frac{dR_o}{dSOC} & -1 \end{bmatrix}$, $D_k = [-R_O(k)]$.

Where x_k and u_k are the system state vectors and the measured system inputs at discrete-time k, and w_k is the unmeasured process noise that can influence the system state. The output of the system is y_k, and v_k is the measurement noise. A_k, C_k are the first partial derivatives matrices with respect to x_k, and B_k, and D_k are the Jacobian matrice with respect to u_k.

With the matrice C_k, $\frac{dV_{oc}}{dSOC} - \frac{dR_o}{dSOC} = M_s$ is a function of the SOC, and thus it can be built from the battery experiment data.

3.3.2. Parameter Identification

In this section, we identify the internal parameters of the built battery model including V_{OC}, R_0, R_p, C_p, and M_s. This requires the experiment data including current and voltage across battery terminals. The V_{OC} of the battery is the potential difference between the poles of the battery in the no load state. In order to obtain open-circuit voltage and the battery SOC curve, the battery was charged in the standard way and then completely discharged with a constant current. The battery was discharged from 100% to 0% in 5% steps, and the period between the two steps is in a no-load (open-circuit) state; this procedure is completed within 2 h to achieve stable V_{OC}. From obtained data, the mathematical relationship between the SOC and V_{OC} was built by using the nonlinear curve fitting, and the fitting results are shown in Equation (9) and Figure 7.

$$V_{OC} = 2.926 + 0.044 \times SOC^1 - 0.0012 \times SOC^2 + 1.511E^{-5} \times SOC^3 - 6.72E^{-8} \times SOC^4 \quad (9)$$

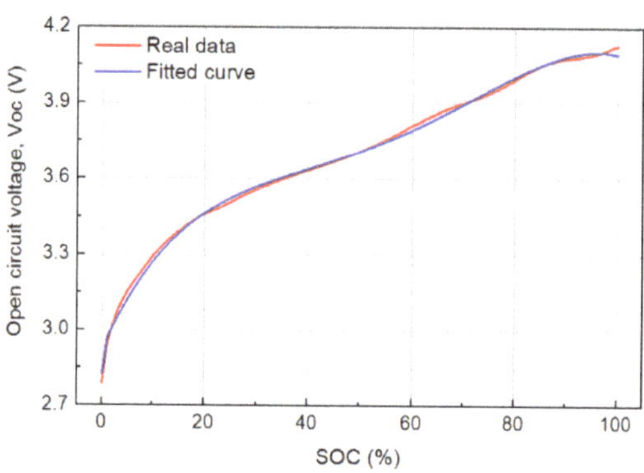

Figure 7. V_{OC} and SOC curve of the LiB.

In the case of R_0, it can also be determined by using the above battery discharge data. The value of R_0 was calculated corresponding to the SOC of each period by dividing the value of the voltage drop by the discharge current, where voltage drop is the difference between the no load and loaded voltages. The variation of R_0 according to the SOC is given in Figure 8, which fits into Equation (10).

$$R_0 = 5.7E^{-3} - 3.4E^{-2}SOC^1 + 0.146SOC^2 - 0.326SOC^3 + 0.415SOC^4 - 0.29SOC^5 + 0.094SOC^6 - 6.4E^{-3}SOC^7 \quad (10)$$

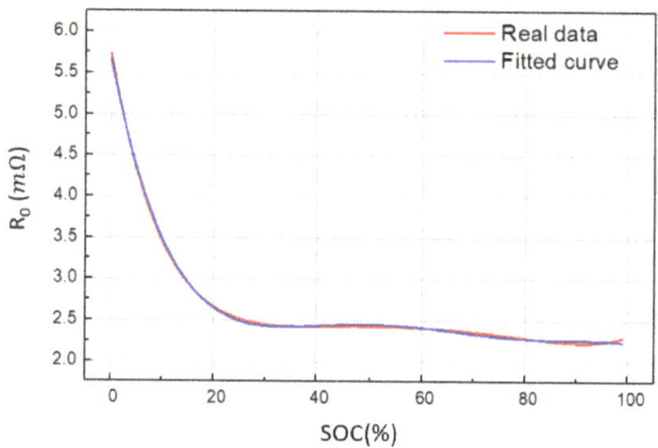

Figure 8. R_0 and SOC curve of the LiB.

From the results of V_{OC} and R_0, we can define the functions of M_s according to the SOC as below.

$$M_s = -14.885+130.43 \times SOC^1 - 588.464 \times SOC^2 + 1381.381 \times SOC^3 - 1889.537 \times SOC^4 \quad (11)$$

In order to identify the remaining battery parameters, R_p and C_p, the Matlab toolbox "Parameter estimate", which is a powerful interface for battery parameter identification, was used. The obtained functions of R_p and C_p according to the SOC are shown in Equations (12) and (13), respectively. Figure 9 describes the variation of R_p and C_p according to the SOC.

$$R_P = 0.015 - 0.18SOC^1 + 1.052SOC^2 - 3.28SOC^3 + 5.798SOC^4 - 5.82SOC^5 + 3.08SOC^6 - 0.67SOC^7 \quad (12)$$

$$C_P = 1.3E3 - 2.8E4 \times SOC + 2.6E5 \times SOC^2 - 1.09E6 \times SOC^3 + 2.4E6 \times SOC^4 - 3.08E6 \times SOC^5 + 2.01E6 \times SOC^6 - 5.3E5 \times SOC^7 \quad (13)$$

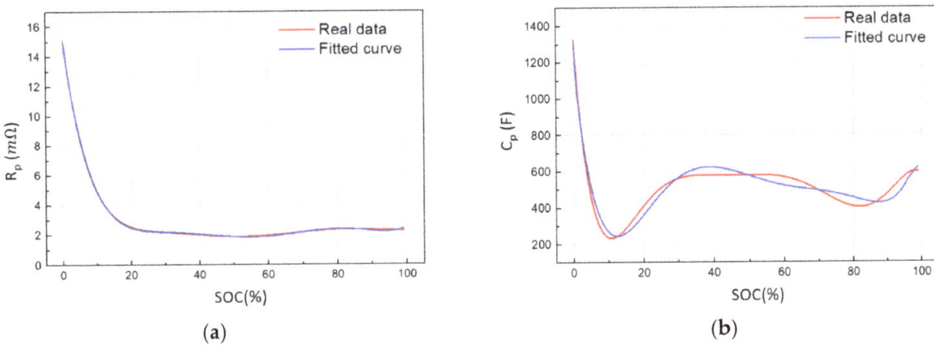

Figure 9. Variation of *Rp* and *Cp* according to SOC: (**a**) *Rp*; and (**b**) *Cp*.

3.3.3. SOC Estimation Model with the EKF

This algorithm combines the Thevenin battery model with the Ah integration method; the SOC and V_p represent the system state variables, the current represents input variables, and the battery terminal voltage represents the output variable. Figure 10 shows the diagram of the SOC estimation method using the EKF.

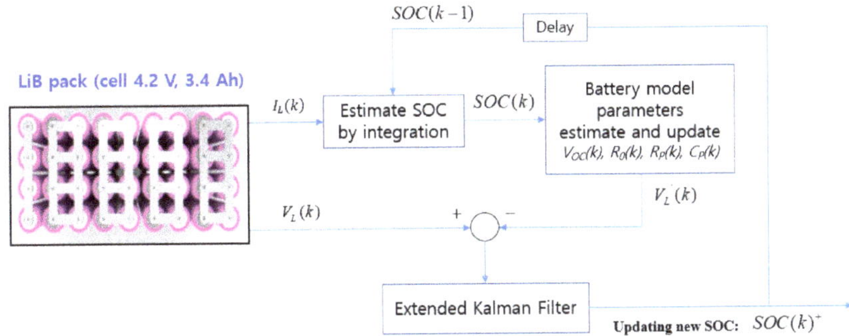

Figure 10. Diagram of the SOC estimation method based on the EKF.

The use of the EKF requires the linearization of the state equations around the operating point for each sample. This algorithm is described in detail as follows:

- Definition of the battery model with the state Equation (8)
- Definition of the noise matrices: $Q = E\{v_k, v_k^T\}$ and $R = E\{w_k, w_k^T\}$
- Initializing the state variable, $x_0^+ = x_0$ and the estimation error $P_0^+ = P_0$
- Prediction of the state variables, $x_k = [SOC(k), V_p(k)]$ at time k by the Equations (6) and (7) from the measured current $I_L(k)$
- Linearization by calculating Jacobians: $F = \left.\frac{\partial f(x_k,u_k)}{\partial x_k}\right|_{x_k^+}$ and $H = \left.\frac{\partial h(x_k,u_k)}{\partial x_k}\right|_{x_k^+}$
- Prediction of the covariance matrix P_k (estimation error): $P_k = F \cdot P_{k-1}^+ \cdot F^T + Q$
- Computation of the Kalman gain, Kg: $K_g = P_k \cdot H^T \cdot (H \cdot P_k \cdot H^T + R)^{-1}$
- Update the new SOC based on the measured voltage, $V_L(k)$ and the output variable $V_L'(k)$:

$$SOC(k)^+ = SOC(k) + K_g \cdot [V(k) - V'(k)] \tag{14}$$

- Update the estimation error, P_k to P_k^+ for the next time step $(k + 1)$: $P_k^+ = (I_L(k) - K_g \cdot H) \cdot P_k$

3.4. Combination of the EKF and ANN for the SOC Estimation

In the case of using the EKF, the battery model was built for the SOC estimation based on the voltage and current characteristics. However, the battery is a dynamic system, and with a long operation time, the internal parameters of the battery model can be changed by other factors such as temperature, load changes, and SOC. These can cause errors in the SOC estimation model, and the battery modeling is very complex when considering them. With the method using the ANN, there were only three inputs including the voltage, current, and temperature of the present sample, and there was no information in the previous sample. This can cause a large dimension and high computation on the network. As the above results, the designed network had 4 hidden layers with 64 neurons in each layer, which means that the weight and bias matrices can take over a large amount of memory in the MCU and make it difficult to handle computation.

To solve the above-mentioned problems, we suggested a SOC estimation method combining the EKF with ANN as shown in Figure 11. The EKF model was applied to determine the battery SOC in the previous sample, $SOC(k-1)$, based on the measured values of the previous voltage, $V_L(k-1)$ and the current, $I_L(k-1)$. This SOC value was used as one input of an ANN to estimate the SOC in the present sample, $SOC(k)$. The ANN was designed with four inputs including the measured voltage, $V_L(k)$, current, $I_L(k)$, and temperature, $T(k)$ at the present sample. The EKF is incorporated with the ANN to adapt to the dynamic environments, and the $SOC(k-1)$ generated by the EKF takes into account battery hysteresis and measurement noise. Thereby, the accuracy and reliability of the SOC estimation can be improved.

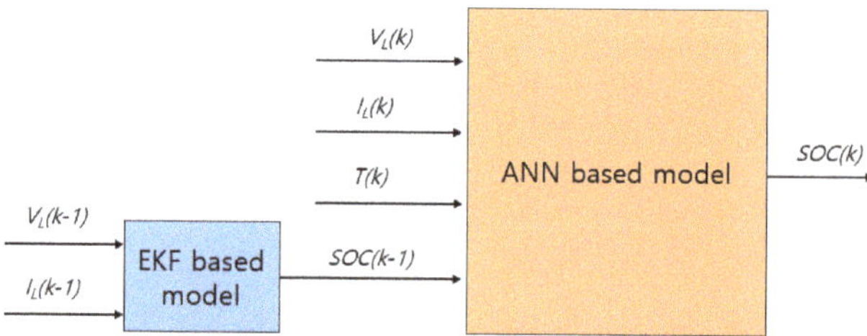

Figure 11. Configuration of the SOC estimation method combining the EKF with ANN.

In this method, the designed EKF model in the previous section was kept to use, and the ANN was designed and trained again with the same battery dataset. The final specifications of the ANN are shown in Table 5. This network only needs two hidden layers, and each hidden layer had 32 neurons. The required memory and computation time in the MCU are significantly reduced. The detailed configuration of the designed ANN including the input layer, hidden layers, output layer, number of neurons in each layer, weight matrices, bias matrices, and activation functions are described in Figure 12.

Table 5. Final specifications of the ANN combined with the EKF.

Specifications	ANN
Number of inputs of the network	4
Number of hidden layers of the network	2
Number of outputs of the network	1
Number of neurons in each hidden layer	32
Activation function in the hidden layers	Sigmoid
Activation function in the output layer	Softplus
Adapting learning function	Adam
Learning rate	0.001
Loss function	MSE
Accuracy metrics	[MSE, MAE]

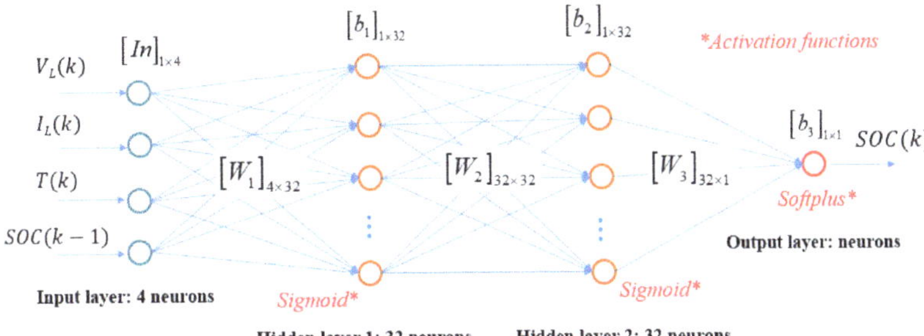

Figure 12. Detailed configuration and parameters of the designed ANN.

The network training was also performed in 500 epochs, and the performances of the redesigned ANN are shown in Figure 13. The MSE and MAE in the final epoch are approximately 0.0064 and 0.057, respectively, which are much lower than that of the ANN using three inputs. The trained model was also tested randomly with 5,000 samples to compare the actual and estimated SOC. The maximum and average errors of the estimated SOC were 0.27% and 0.069%, respectively, as shown in Figure 14.

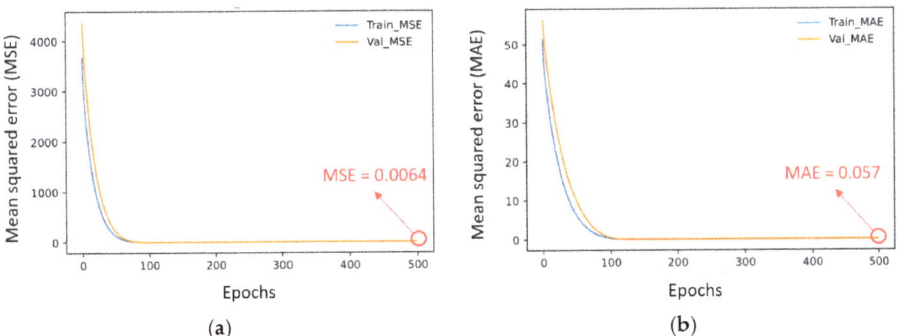

Figure 13. Training performances of the redesigned ANN: (**a**) MSE, and (**b**) MAE.

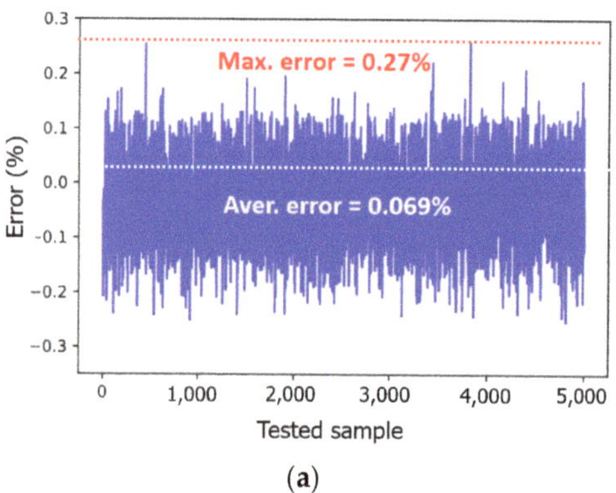

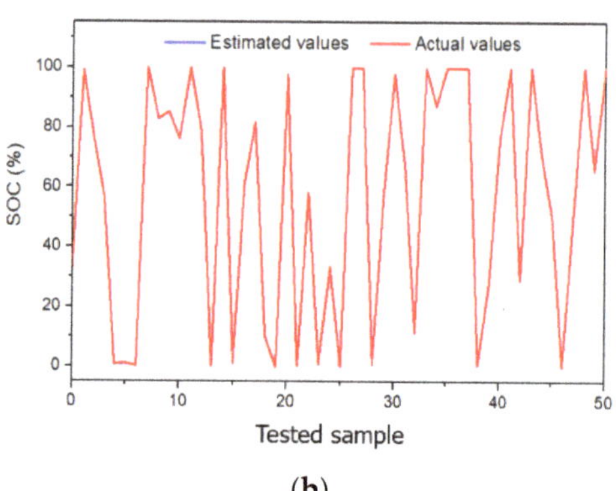

(a) (b)

Figure 14. Test results of the redesigned ANN with 5,000 random samples: (**a**) error of estimated SOC, and (**b**) estimated and actual SOCs.

4. Experiment Results and Discussions

4.1. Implementations of the SOC Estimation Methods in the Smart BMS

To validate the proposed SOC estimation methods, we applied them to the SOC estimation function of the BMS in an experimental test with a real LiB pack. First, we implemented a real-time SOC estimation based on the ANN. Then, the SOC estimations based on the EKF and the EKF-ANN were implemented in offline experiments by using the MATLAB program. These models were built to calculate the SOC from the experiment data acquired using the ANN. Finally, the obtained results of three methods were compared with the reference SOC, which was calculated from the measured battery capacity.

Using the ANN, the SOC was estimated based on the average voltage and current of the battery cells and the temperature in the battery pack. The SOC estimation function was built by C programming language on the master MCU STM32F205, which has a flash memory of up to 1 Mbyte. In the designed ANN, there were three 64 × 64 weight matrices and four 1x64 bias matrices in the hidden layers, and it took about 24 ms for calculation speed. The SOC was calculated in each sampling time of 1 s with the full charge and discharge of the battery pack. We found no problems with the memory and operation of the master MCU during the experiment. A monitoring system was also facilitated for the BMS as shown in Figure 15. The main interface of the monitoring system includes the contents listed below:

(1) Configuration of the BMS
(2) Measurement values including SOC, local time, and BMS version
(3) Status of the BMS including warming, detection, state, and status
(4) Summary of main measurement values
(5) Serial communications with the BMS
(6) Command to read and write the alarm and cut-off values
(7) Calibration of the pack voltage and current values
(8) External inputs to block the relay
(9) Main screen displaying measurement values in the master and slave MCUs

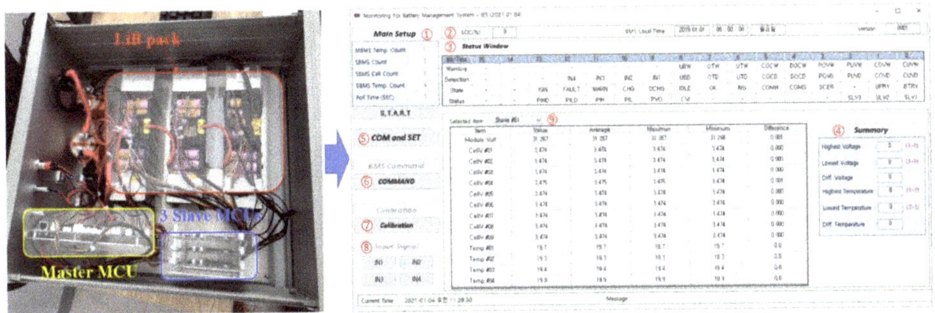

Figure 15. Monitoring system for the smart BMS.

4.2. Experiment Results of the SOC Estimation Methods

Since the proposed SOC estimation methods were built based on the battery data, these methods only valid under the experimental conditions similar to the range of battery data collected as follows:

- After measuring the voltage and current of the entire battery pack, calculate the average value of the voltage and current of one cell,
- Battery pack is discharged and charged with a constant current of 1.7 A in each cell,
- Cell voltage range is 2.6–4.2 V,
- Temperature range of the battery pack is 19~42 °C.

In the experiment, we first performed the SOC estimation using the ANN in real-time. The average voltage and current of the cells and the measured temperature during the discharge and charge processes of the battery pack are described in Figure 16. The battery pack was fully discharged and charged with the same currents of 1.7 A in each cell, which were similar to the trained battery dataset. The temperature was measured at the center of the battery pack, and the temperature range was from 19 °C to 38 °C.

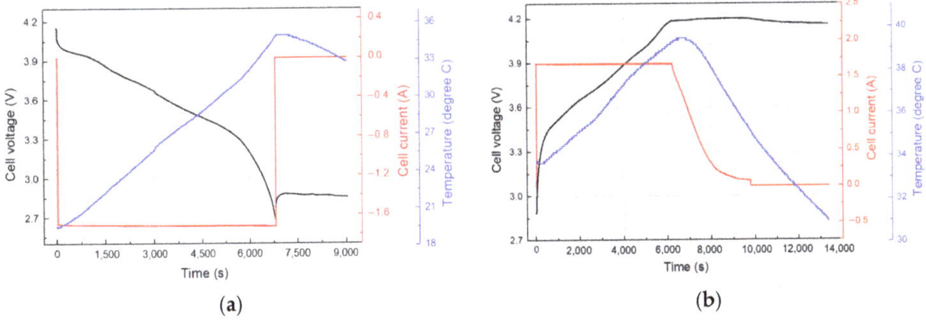

Figure 16. Experiment results for the battery pack characteristics: (a) discharge; and (b) charge.

Figure 17 shows the experimental results of the online SOC estimation using the ANN, and the estimated SOC was compared with the reference SOC. As a result, the maximum SOC errors in the discharge and charge processes were 2.3% and 2.6%, respectively. This result has satisfied the initial design target of the SOC estimation error for the smart BMS.

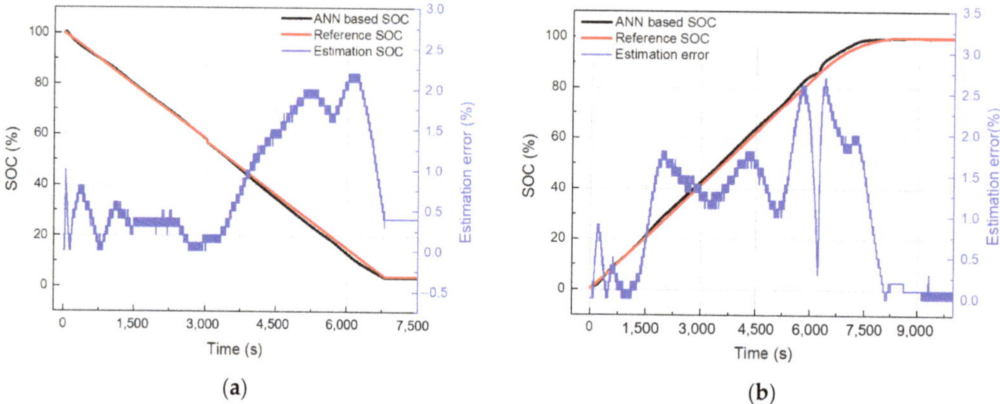

Figure 17. Experiment results of the SOC estimation using the ANN: (**a**) discharge; and (**b**) charge.

Next, the above experiment data including the battery voltage, current, and temperature were used to calculate the SOC offline using the other two methods in the Matlab simulation model. Figure 18 shows the comparison of the SOC estimation results using three methods with the reference SOC during the discharge and charge processes of the battery. The detailed absolute SOC errors of each method are given in Figure 19 and Table 6. Using the EKF, the maximum SOC errors in the discharge and charge processes were 2.8% and 2.4%, respectively, which were similar to that of using the ANN. We found the significant improvement for the SOC estimation by combining the EKF with ANN, which had the SOC error of less than 1%. Comparisons were made with other SOC estimation methods [37–49] and the maximum estimation errors are summarized in Table 7. Through this, it was confirmed that the proposed method guarantees the accuracy of SOC estimation similar to or better than other methods.

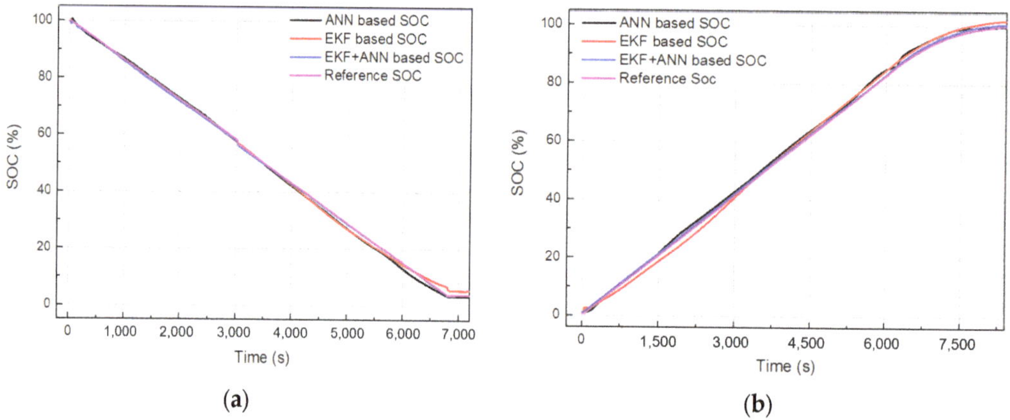

Figure 18. Experiment results of the SOC estimations using three methods: (**a**) Discharge; and (**b**) Charge.

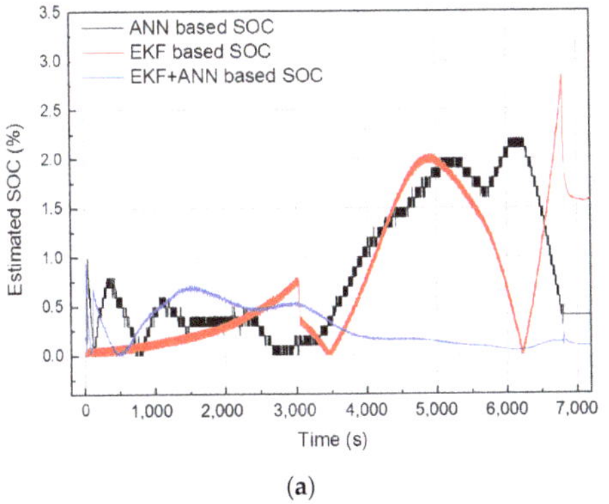

(**a**)

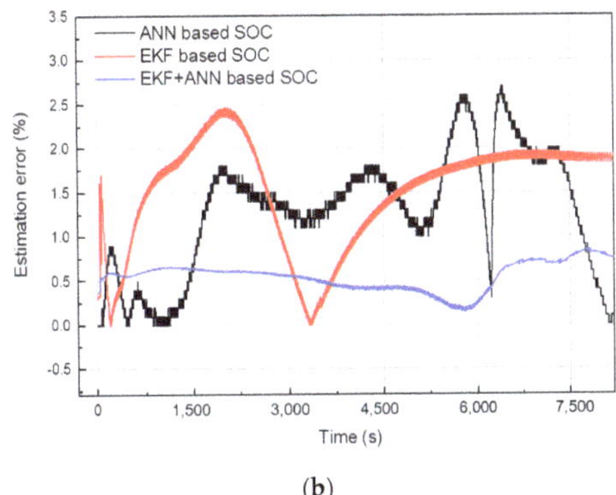

(**b**)

Figure 19. Errors of the SOC estimations using three methods: (**a**) Discharge; and (**b**) charge.

Table 6. Summary of the errors in the SOC estimations using three methods.

Methods		ANN Based	EKF Based	EKF+ANN Based
Discharge	Maximum error	2.3%	2.8%	0.9%
	Average error	0.86%	0.77%	0.29%
Charge	Maximum error	2.6%	2.4%	0.8%
	Average error	1.26%	1.53%	0.55%

Table 7. Comparison of the designed SOC estimation methods with others.

SOC Estimation Methods		Maximum Estimation Errors
This study	ANN based	2.6%
	EKF based	2.8%
	EKF+ANN based	0.9%
Other studies	Fuzzy logic based	2.12% [37], 2.45% [38], 1.9% [39]
	Impedance based	4% [40], 1% [41], 2.3% [42]
	Unscented Kalman Filter based	1.22% [43], 1.5% [44]
	Adaptive EKF based	2% [45], 2.5% [46], 1.6% [47]
	Coulomb counting	3% [48], 4% [49]

4.3. Discussions

From the above experimental results, it can be seen that the SOC estimation methods proposed in this study were accurate and satisfied the requirements of the designed smart BMS. However, there are still issues to be investigated. The performances of the method were evaluated when the battery pack was discharged and charged with a constant current. In a real battery system such as an EV, the load current continuously changes according to the vehicle speed. Therefore, to improve the quality of the battery modeling in the EKF method, the dynamic characteristics of the battery must be considered. Applying an ANN to EV requires more battery data trained with a dynamic current profile. In addition, other important issues of SOC estimation and BMS design, such as cell balancing and battery capacity fade, have not yet been considered. During long-term operation, more experiments need to be performed to collect battery data for each cell and analyze the cathode chemistry of the cells entirely. The next study will consider these issues and improve the accuracy of the SOC estimation method for the LiBs in various real systems.

5. Conclusions

The authors proposed effective SOC estimation methods based on the EKF and ANN for a LiB pack in a smart BMS. Detailed configurations and specifications of the smart BMS and LiB pack were presented. First, an ANN was used to build an SOC estimation model, which was trained and tested using a real battery dataset including voltage, current, temperature, and measured SOC over 20 cycles. Inputs of the designed network consisted of voltage, current, and temperature, and output was the estimated SOC of the battery. The design process for the ANN was described in detail. The Google TensorFlow open-source library was used to design and optimize the network configurations. Next, we developed a SOC estimation algorithm using the EKF, in which the LiB model was studied and a Thevenin model was developed to combine it with the Ah integration method. The current and terminal voltage of the battery represent the input variables, and the SOC represents the output variable. Finally, the EKF-ANN was proposed to improve the shortcomings of the above two methods, where the ANN was redesigned by adding one more input of the previous SOC determined using the EKF method. Both methods were confirmed through experiments performed on real battery data collected from the battery pack consisting of the LIB 18,650 35E cells at 4.2 V and 3.4 Ah. With the ANN and EKF, the SOC estimation performances were almost similar with a maximum SOC errors of 2.4% to 2.8%. Meanwhile, the use of the EKF-ANN significantly improved the accuracy of SOC estimation with less than 1% error. We are confident that the results of this study can be effectively applied to a smart BMS for industrial energy storage systems.

Author Contributions: Conceptualization, methodology: V.Q.D., M.-C.D. and C.S.K.; writing-original draft preparation: V.Q.D.; experiment support: B.Z., C.-H.D., L.J. and M.-K.L.; validation, writing-review and editing: M.-C.D. and B.Z.; project administration: J.H.B.; supervision: M.P. All authors have read and agreed to the published version of the manuscript.

Funding: This research was supported by the "Development of high precision Smart-BMS for fire prevention with HKF and Deep Learning" project of IES Co., LTD (Ho Chi Minh City, Vietnam) and Korea Electrotechnology Research Institute.

Institutional Review Board Statement: Not applicable.

Informed Consent Statement: Not applicable.

Data Availability Statement: Data is contained within the article.

Conflicts of Interest: The authors declare no conflict of interest.

References

1. Khalid, M.; Sheikh, S.S.; Janjua, A.K.; Khalid, H.A. Performance validation of electric vehicle's battery management system under state of charge estimation for lithium-ion battery. In Proceedings of the 2018 International Conference on Computing, Electronic and Electrical Engineering (ICE Cube), Quetta, Pakistan, 12–13 November 2018; pp. 1–5. [CrossRef]
2. Zheng, M.X.; Qi, B.J.; Du, X.W. Dynamic model for characteristics of Li-ion battery on electric vehicle. In Proceedings of the 2009 4th IEEE Conference on Industrial Electronics and Applications, Xian, China, 25–27 May 2009; pp. 2867–2871. [CrossRef]
3. Chen, T.M.; Jin, Y.; Lv, H.Y.; Yang, A.T.; Liu, M.Y.; Chen, B.; Xie, Y.; Chen, Y. *Applications of Lithium-Ion Batteries in Grid-Scale Energy Storage Systems*; Transactions of Tianjin University: Tianjin, China, 2020; Volume 26, pp. 208–217. [CrossRef]
4. Mehr, T.H.; Masoum, M.A.S.; Jabalameli, N. Grid-connected Lithium-ion battery energy storage system for load leveling and peak shaving. In Proceedings of the 2013 Australasian Universities Power Engineering Conference (AUPEC), Hobart, Australasia, 29 September 2013; pp. 1–6. [CrossRef]
5. Thorbergsson, E.; Knap, V.; Swierczynski, M.; Stroe, D.; Teodorescu, R. Primary Frequency Regulation with Li-Ion Battery Based Energy Storage System-Evaluation and Comparison of Different Control Strategies. In Proceedings of the Intelec 2013 35th International Telecommunications Energy Conference, Smart Power and Efficiency, Hamburg, Germany, 13–17 October 2013; pp. 1–6.
6. Byrne, R.H.; Hamilton, S.; Borneo, D.R.; Olinsky-Paul, T.; Gyuk, I. The value proposition for energy storage at the sterling municipal light department. In Proceedings of the 2017 IEEE Power & Energy Society General Meeting, Chicago, IL, USA, 16–20 July 2017; pp. 1–5. [CrossRef]
7. Hannan, M.A.; Hoque, M.M.; Hussain, A.; Yusof, Y.; Ker, P.J. *State-of-the-Art and Energy Management System of Lithium-Ion Batteries in Electric Vehicle Applications: Issues and Recommendations*; IEEE: Piscataway, NJ, USA, 2018; Volume 6, pp. 19362–19378. [CrossRef]

8. Duan, J.; Tang, X.; Dai, H.; Yang, Y.; Wu, W.; Wei, X.; Huang, Y. Building Safe Lithium-Ion Batteries for Electric Vehicles: A Review. *Electrochem. Energ. Rev.* **2020**, *3*, 1–42. [CrossRef]
9. How, D.N.T.; Hannan, M.A.; Hossain Lipu, M.S.; Ker, P.J. *State of Charge Estimation for Lithium-Ion Batteries Using Model-Based and Data-Driven Methods: A Review*; IEEE: Piscataway, NJ, USA, 2019; Volume 7, pp. 136116–136136. [CrossRef]
10. Naguib, M.; Kollmeyer, P.; Emadi, A. *Lithium-Ion Battery Pack Robust State of Charge Estimation, Cell Inconsistency, and Balancing: Review*; IEEE: Piscataway, NJ, USA, 2021. [CrossRef]
11. Mohamed, A.T. A review paper on batteries charging systems with the state of charge determination techniques. In Proceedings of the 2nd Smart Cities Symposium (SCS 2019), Bahrain, Bahrain, 24–26 March 2019; pp. 1–6. [CrossRef]
12. Yuan, S.; Wu, H.; Yin, C. State of Charge Estimation Using the Extended Kalman Filter for Battery Management Systems Based on the ARX Battery Model. *Energies* **2013**, *6*, 444–470. [CrossRef]
13. He, H.; Xiong, R.; Zhang, X.; Sun, F.; Fan, J. State-of-Charge Estimation of the Lithium-Ion Battery Using an Adaptive Extended Kalman Filter Based on an Improved Thevenin Model. *IEEE Trans. Veh. Technol.* **2011**, *60*, 1461–1469. [CrossRef]
14. Chitnis, M.S.; Pandit, S.P.; Shaikh, M.N. Electric Vehicle Li-Ion Battery State of Charge Estimation Using Artificial Neural Network. In Proceedings of the 2018 International Conference on Inventive Research in Computing Applications (ICIRCA), Coimbatore, India, 11–12 July 2018; pp. 992–995. [CrossRef]
15. Xiong, R.; Cao, J.; Yu, Q.; He, H.; Sun, F. *Critical Review on the Battery State of Charge Estimation Methods for Electric Vehicles*; IEEE: Piscataway, NJ, USA, 2018; Volume 6, pp. 1832–1843. [CrossRef]
16. Manthopoulos, A.; Wang, X. A Review and Comparison of Lithium-Ion Battery SOC Estimation Methods for Electric Vehicles. In Proceedings of the IECON 2020 The 46th Annual Conference of the IEEE Industrial Electronics Society, Singapore, 18–21 October 2020; pp. 2385–2392. [CrossRef]
17. Chang, W.Y. The State of Charge Estimating Methods for Battery: A Review. *Int. Sch. Res. Not.* **2013**, *2013*, 953792. [CrossRef]
18. Zhang, R.; Xia, B.; Li, B.; Cao, L.; Lai, Y.; Zheng, W.; Wang, H.; Wang, W.; Wang, M. A Study on the Open Circuit Voltage and State of Charge Characterization of High-Capacity Lithium-Ion Battery under Different Temperature. *Energies* **2018**, *11*, 2408. [CrossRef]
19. Nugroho, A.; Rijanto, E.; Wijaya, F.D.; Nugroho, P. Battery state of charge estimation by using a combination of Coulomb Counting and dynamic model with adjusted gain. In Proceedings of the 2015 International Conference on Sustainable Energy Engineering and Application (ICSEEA), Bandung, Indonesia, 5–7 October 2015; pp. 54–58. [CrossRef]
20. Cuadras, A.; Kanoun, O. SoC Li-ion battery monitoring with impedance spectroscopy. In Proceedings of the 2009 6th International Multi-Conference on Systems, Signals and Devices, Djerba, Tunisia, 23–26 March 2009; pp. 1–5. [CrossRef]
21. Spagnol, P.; Rossi, S.; Savaresi, S.M. Kalman Filter SoC estimation for Li-Ion batteries. In Proceedings of the2011 IEEE International Conference on Control Applications (CCA), Denver, CO, USA, 28–30 September 2011; pp. 587–592. [CrossRef]
22. Rahmoun, A.; Biechl, H.; Rosin, A. SOC estimation for Li-Ion batteries based on equivalent circuit diagrams and the application of a Kalman filter. In Proceedings of the 2012 Electric Power Quality and Supply Reliability, Tartu, Estonia, 11–13 June 2012; pp. 1–4. [CrossRef]
23. Lee, S.J.; Kim, J.H.; Lee, J.M.; Cho, B.H. The State and Parameter Estimation of an Li-Ion Battery Using a New OCV-SOC Concept. In Proceedings of the 2007 IEEE Power Electronics Specialists Conference, Orlando, FL, USA, 17–21 June 2007; pp. 2799–2803. [CrossRef]
24. Yatsui, M.W.; Bai, H. Kalman filter based state-of-charge estimation for lithium-ion batteries in hybrid electric vehicles using pulse charging. In Proceedings of the 2011 IEEE Vehicle Power and Propulsion Conference, Chicago, IL, USA, 6–9 September 2011; pp. 1–5. [CrossRef]
25. Haoran, L.; Liangdong, L.; Xiaoyin, Z.; Mingxuan, S. Lithium Battery SOC Estimation Based on Extended Kalman Filtering Algorithm. In Proceedings of the 2018 IEEE 4th International Conference on Control Science and Systems Engineering (ICCSSE), Wuhan, China, 21–23 August 2018; pp. 231–235. [CrossRef]
26. Yang, S.; Zhou, S.; Hua, Y.; Zhou, X.A.; Liu, X.H.; Pan, Y.W.; Ling, H.P.; Wu, B. A parameter adaptive method for state of charge estimation of lithium-ion batteries with an improved extended Kalman filter. *Sci. Rep.* **2021**, *11*, 5805. [CrossRef] [PubMed]
27. Ismail, M.; Dlyma, R.; Elrakaybi, A.; Ahmed, R.; Habibi, S. Battery state of charge estimation using an Artificial Neural Network. In Proceedings of the 2017 IEEE Transportation Electrification Conference and Expo (ITEC), Chicago, IL, USA, 22–24 June 2017; pp. 342–349. [CrossRef]
28. Almeida, G.C.S.; Souza, A.C.Z.d.; Ribeiro, P.F. A Neural Network Application for a Lithium-Ion Battery Pack State-of-Charge Estimator with Enhanced Accuracy. *Proceedings* **2020**, *58*, 33. [CrossRef]
29. Saji, D.; Babu, P.S.; Ilango, K. SoC Estimation of Lithium-Ion Battery Using Combined Coulomb Counting and Fuzzy Logic Method. In Proceedings of the 2019 4th International Conference on Recent Trends on Electronics, Information, Communication & Technology (RTEICT), Bangalore, India, 17–18 May 2019; pp. 948–952. [CrossRef]
30. Rivera-Barrera, J.P.; Muñoz-Galeano, N.; Sarmiento-Maldonado, H.O. SoC Estimation for Lithium-ion Batteries: Review and Future Challenges. *Electronics* **2017**, *6*, 102. [CrossRef]
31. Liu, K.; Li, K.; Peng, Q.; Zhang, C. A brief review on key technologies in the battery management system of electric vehicles. Front. *Mech. Eng.* **2019**, *14*, 47–64. [CrossRef]

32. Vaideeswaran, V.; Bhuvanesh, S.; Devasena, M. Battery Management Systems for Electric Vehicles using Lithium-Ion Batteries. In Proceedings of the 2019 Innovations in Power and Advanced Computing Technologies (i-PACT), Vellore, India, 22–23 March 2019; pp. 1–9. [CrossRef]
33. Ali, S.S.; Choi, B.J. State-of-the-Art Artificial Intelligence Techniques for Distributed Smart Grids: A Review. *Electronics* **2020**, *9*, 1030. [CrossRef]
34. Lai, X.; Qin, C.; Gao, W.; Zheng, Y.; Yi, W. A State of Charge Estimator Based Extended Kalman Filter Using an Electrochemistry-Based Equivalent Circuit Model for Lithium-Ion Batteries. *Appl. Sci.* **2018**, *8*, 1592. [CrossRef]
35. He, H.; Xiong, R.; Fan, J. Evaluation of Lithium-Ion Battery Equivalent Circuit Models for State of Charge Estimation by an Experimental Approach. *Energies* **2011**, *4*, 582–598. [CrossRef]
36. Chen, M.; Rincon-Mora, G.A. *Accurate Electrical Battery Model Capable of Predicting Runtime and I–V Performance*; IEEE Transactions on Energy Conversion: Piscataway, NJ, USA, 2006; Volume 21, pp. 504–511. [CrossRef]
37. Jiani, D.; Zhitao, L.; Youyi, W.; Changyun, W. A fuzzy logic-based model for Li-ion battery with SOC and temperature effect. In Proceedings of the 11th IEEE International Conference on Control & Automation (ICCA), Taichung, Taiwan, 18–20 June 2014; pp. 1333–1338. [CrossRef]
38. Zheng, W.; Xia, B.; Wang, W.; Lai, Y.; Wang, M.; Wang, H. State of Charge Estimation for Power Lithium-Ion Battery Using a Fuzzy Logic Sliding Mode Observer. *Energies* **2019**, *12*, 2491. [CrossRef]
39. Rao, R.P.; Bhat, R.S.; Ranjeeth, R.; Kavya, B.G. Implementing Fuzzy Logic to Improve the Accuracy of SoC Estimation for Li-ion Battery. *Int. J. Eng. Res. Technol.* **2020**, *09*, 937–941.
40. Wu, S.; Chen, H.; Tsai, M.; Lin, T.; Chen, L. AC Impedance Based Online State-of-Charge Estimation for Li-ion Battery. In Proceedings of the 2017 International Conference on Information, Communication and Engineering (ICICE), Xiamen, China, 17–20 November 2017; pp. 53–56. [CrossRef]
41. Xu, J.; Chris Mi, C.; Cao, B.G.; Cao, J.Y. A new method to estimate the state of charge of lithium-ion batteries based on the battery impedance model. *J. Power Sources* **2013**, *233*, 277–284. [CrossRef]
42. Kim, T.; Qiao, W.; Qu, L. Real-time state of charge and electrical impedance estimation for lithium-ion batteries based on a hybrid battery model. In Proceedings of the 2013 Twenty-Eighth Annual IEEE Applied Power Electronics Conference and Exposition (APEC), Long Beach, CA, USA, 17–21 March 2013; pp. 563–568. [CrossRef]
43. Guo, X.; Xu, X.; Geng, J.; Hua, X.; Gao, Y.; Liu, Z. SOC Estimation with an Adaptive Unscented Kalman Filter Based on Model Parameter Optimization. *Appl. Sci.* **2019**, *9*, 4177. [CrossRef]
44. Wang, W.; Wang, X.T.; Xiang, C.; Wei, C.; Zhao, Y. *Unscented Kalman Filter-Based Battery SOC Estimation and Peak Power Prediction Method for Power Distribution of Hybrid Electric Vehicles*; IEEE: Piscataway, NJ, USA, 2018; Volume 6, pp. 35957–35965. [CrossRef]
45. Mazzi, Y.; Sassi, H.B.; Errahimi, F.; Es-Sbai, N. State of charge estimation using extended kalman filter. In Proceedings of the 2019 International Conference on Wireless Technologies, Embedded and Intelligent Systems (WITS), Fez, Morocco, 3–4 April 2019; pp. 1–6. [CrossRef]
46. Ciortea, F.; Rusu, C.; Nemes, M.; Gatea, C. Extended Kalman Filter for state-of-charge estimation in electric vehicles battery packs. In Proceedings of the 2017 International Conference on Optimization of Electrical and Electronic Equipment (OPTIM) & 2017 Intl Aegean Conference on Electrical Machines and Power Electronics (ACEMP), Brasov, Romania, 25–27 May 2017; pp. 611–616. [CrossRef]
47. Cheng, Z.; Lv, J.K.; Liu, Y.L.; Yan, Z.H. Estimation of State of Charge for Lithium-Ion Battery Based on Finite Difference Extended Kalman Filter. *J. Appl. Math.* **2014**, *2014*, 10. [CrossRef]
48. Baccouche, I.; Mlayah, A.; Jemmali, S.; Manai, B.; Essoukri Ben Amara, N. $Implementation of a Coulomb counting algorithm for SOC estimation of Li-Ion battery for multimedia applications. In Proceedings of the 2015 IEEE 12th International Multi-Conference on Systems, Signals & Devices (SSD15), Mahdia, Tunisia, 16–19 March 2015; pp. 1–6. [CrossRef]
49. Qaisar, S.M. Event-Driven Coulomb Counting for Effective Online Approximation of Li-Ion Battery State of Charge. *Energies* **2020**, *13*, 5600. [CrossRef]

Article

Simulation of Energy Exchange between Single Prosumer Residential Building and Utility Grid

Andres Annuk [1,*], Wahiba Yaïci [2], Matti Lehtonen [3], Risto Ilves [4], Toivo Kabanen [1] and Peep Miidla [5]

1 Department of Energy Application Engineering, Institute of Technology, Estonian University of Life Sciences, 51006 Tartu, Estonia; toivo.kabanen@emu.ee
2 CanmetENERGY Research Centre, Natural Resources Canada, 1 Haanel Drive, Ottawa, ON K1A 1M1, Canada; wahiba.yaici@canada.ca
3 Department of Electrical Engineering and Automation, Aalto University, 02150 Espoo, Finland; matti.lehtonen@aalto.fi
4 Department of Biosystems Engineering, Institute of Technology, Estonian University of Life Sciences, 51006 Tartu, Estonia; risto.ilves@emu.ee
5 Estonian Center of Industrial Mathematics, 50090 Tartu, Estonia; peep.miidla@ut.ee
* Correspondence: andres.annuk@emu.ee; Tel.: +372-55-6820624

Abstract: Modern households usually have independent energy sources such as wind generators, photovoltaic (PV) panels, and similar green energy production equipment. Experts predict that soon, there will be an increasing number of such prosumers who both produce and consume energy. This process alleviates and reduces the load on large national electricity networks and also contributes to overall energy security. In this paper, a simulation model of a household, which employs a wind generator as its independent source of electricity, is developed. It is expected that this approach will be easily replicated for more complex configurations. The other components of the single prosumer microgrid that will be assessed are the non-shiftable electricity consumption equipment, which is used mainly in households and deployed separately for water heater, with a separate battery to meet the needs of these non-shiftable consumers. The 5-min data intervals for the year of simulation have been used. The characteristics of energy flow according to production and consumption schedules and the capacity of storage equipment have been modelled and simulated. Results disclose that wind turbine production size and buffer battery have a crucial impact on the demand cover factor.

Keywords: load shifting; energy storage; wind energy; green energy; self-consumption; cover factor; microgrids; buffer battery; distributed generation; simulation

Citation: Annuk, A.; Yaïci, W.; Lehtonen, M.; Ilves, R.; Kabanen, T.; Miidla, P. Simulation of Energy Exchange between Single Prosumer Residential Building and Utility Grid. *Energies* **2021**, *14*, 1553. https://doi.org/10.3390/en14061553

Academic Editor: Andrei Blinov

Received: 10 February 2021
Accepted: 8 March 2021
Published: 11 March 2021

Publisher's Note: MDPI stays neutral with regard to jurisdictional claims in published maps and institutional affiliations.

1. Introduction

The European Union's energy policy aims to achieve 32% of its total electricity production from renewable sources by 2030 [1]. Increasingly, more attention is being paid to converting energy from renewable sources such as wind and solar [2]. These natural energy sources present new challenges for electrical engineers and researchers regarding maximal extractions of energy for on-site use, and the methods of their deployment besides fossil energy sources [3]. The renewable sources are interfaced with the grid by the means of power electronic converters. Such systems are becoming widespread and getting more efficient with the developments in topologies and power semiconductor components [4,5]. In terms of national energy security, it is important to maintain enough rotating reserve because this helps to maintain the stability of the electricity grid [6,7]. Small cogeneration heat plants that are powered by biofuels can support grid stability to an extent, but the combustion of biofuels on the other hand harms the environment [8]. Hydropower is environmentally friendly and stable in 24-h cycles, although it varies from season to season [9,10].

Another important aspect of electricity use is storage. Storage is possible only at varying levels. Pumped hydropower plants can be used for large-scale electricity storage [11].

Flywheel storage [12] is also used as a storage device in smart grids, transportation and for maintaining grid stability; however, for the most part, private use remains theoretical. Europe largely supports the transition to minimize energy use in which most new buildings will consume nearly zero energy with autonomous energy production and adequate storage systems. In the past decade, microgrid-based prosumers have grown exponentially, and it is desired that these prosumers would play a more appreciable role in optimizing the operations of utility grids [13–15].

This article provides a simulation model of a household with a wind generator as its electricity source. The other components of microgrid under review are so-called non-shiftable equipment for electricity consumption, which are used in every home. We consider these appliances and household equipment, known as "non-shiftable" (NS) consumption, in the energy scheme and simulation experiments. Water consumption and water tank as storage are evaluated separately in the model because of the requirement for domestic water to be preheated before use. The final component of the energy scheme within the microgrid or household is a separate battery that ensures that the needs of NS consumers are best met and which can store as much energy as possible when it is produced by the wind generator.

Conventionally, energy consumed in households emanates from either the wind generator [16] or the utility grid. To attain self-sufficiency, i.e., where electricity generated locally matches local consumption, it is necessary to find an optimal combination of installed electricity production units, and storage buffers to mitigate the volatility of the primary energy carrier such as wind or solar irradiation [2]. The present study introduces a buffer battery (BB), which stands between the microgrid and utility grid. BB belongs to the household as material equipment but is virtually considered as energy, being one of the stores for the energy produced by the wind generator. The energy from the wind generator that is not consumed in the microgrid during the observation period is loaded there. If necessary, the energy needed for consumption in the microgrid is also provided from BB. The energy that cannot be contained in the BB is transmitted to the utility grid. It is, thus, possible that the energy consumed in the household can come from either wind generator, buffer or grid.

Cover factor augmentation is indirectly assessed by certain authors. An overview [17] describes loss minimization and power quality in distributed grids and sets as objective the decrease in active losses in batteries. Vanhoudt et al. [18] studied the possibility of increasing self-consumption by heat pump, which is indirectly related to energy storage. By comparing different renewable electricity sources, they found that the wind generator's yield is better compared to photovoltaic (PV) panels. Naval et al. [2] modelled the versatility of electricity sources and related real-time electricity prices, with the wind as one of the most suitable primary energy carriers. The combination of wind and solar generation in a microgrid was studied in [17,19]. For a net-zero energy (NZE) hybrid microgrid, combined wind/solar generation with intermediate storage was analyzed using the HOMER Pro software (Homer Energy LLC, Boulder, USA))suite [20]. A major disadvantage of this software is the 1-h or longer averaging period that it requires. Therefore, a less granular, self-developed MATLAB (The MathWorks, Inc., Kista, Sweden)model with an averaging period of 5 min was applied in this current research.

The cover factor is an indicator of load-shifting technology and is meant to handle volatile primary energy carriers such as wind and sunlight, by deploying intermediate storage devices [21–26]. In addition to shifting, Eltanay et al. [23] prioritized loads by dividing them into two major groups. In some sources, it is described as load matching index (*LMI*) or load generation matching index (*LGMI*) [27,28]. In [29] the optimal storage capacity for full ride-throughs was discussed. Households require power supplies from the largest possible number of renewable sources to reduce payoff times [2,30]. Increasing self-consumption not only levels load peaks [31] but also decreases costs on the electrical energy import to an economically feasible point of 60% self-consumption level [32].

The numerous studies described above indicate the increasing attention to the subject of the microgrid. However, despite this proliferation of studies trying to solve and predict the energy exchange between prosumer residential building and utility grid, to the authors' knowledge, there is still insufficient research into or investigation on optimal system design of a microgrid system in terms of wind turbine and buffer battery sizes effects on the system performance and demand cover factor. The main novelty of this work is using a buffer battery between of microgrid and utility grid to increasing of self-consumption. It is not yet met in the scientific literature.

Hence, the main goal of this work is to simulate and establish setup configurations of BB between microgrid and utility grid to increase self-consumption of the prosumer. It is important to note that for evaluation of effectiveness and distinguishing of these configurations, a new cover factor was introduced and used. The microgrid parameters were also set. The approach can easily be generalized and replicated in more complex configurations. Modelling and numerical experiments were carried out in a MATLAB environment.

With these objectives in mind, the remaining article is structured as follows: Section 2 describes the configuration of the modelled household. Section 3 gives an overview of the initial data that are used in simulation experiments. Section 4 describes the simulation setup, while Section 5 presents the main results. Finally, the main conclusions drawn in this work are provided in Section 6.

2. System Setup

On the base of our computer simulation is a typical private household with an additional buffer battery (BB) between the local microgrid and the utility grid. The term microgrid refers to an electrical installation, which comprises local electricity generation, loads, storage and utility grid connection by default [33].

In Figure 1, the system setup consists of two scenarios. In the first, the microgrid is connected to the utility grid Figure 1a and the energy change between microgrid (MG) and utility grid (UG) is direct and bidirectional. In Figure 1a, Arrow A denotes energy acquired from UG, while Arrow B means excess energy produced by wind generator (WG) and sold to UG. In the second scenario, the microgrid has external buffer battery (BB) storage attached Figure 1b. In Figure 1b, Arrow C means the most expensive energy acquired from UG for a house owner. Arrow D denotes energy, which is taken back to MG and this is free for a homeowner as it was earlier saved to BB (Arrow E), produced by a wind generator, and leftover from household consumption. Arrow F indicates the possibility that wind generator works well and some energy goes to UG. Figure 2 presents MG. All processes are driven by the load controller.

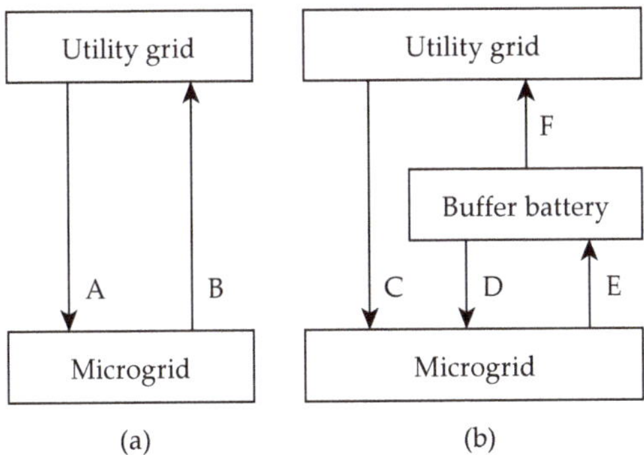

Figure 1. Energy flows between microgrid (MG) and utility grid (UG). (**a**) MG without BB, (**b**) MG with (BB).

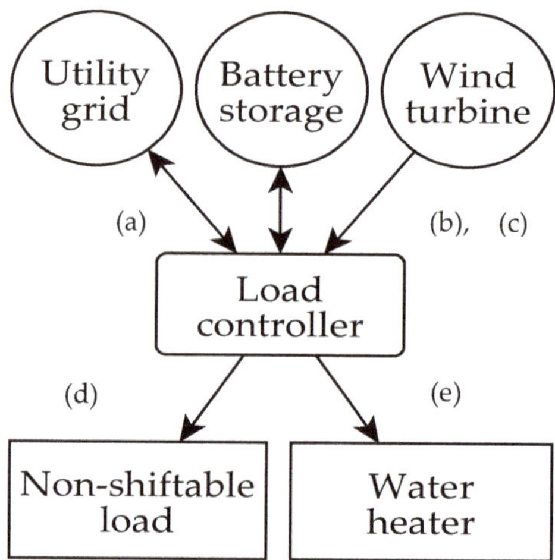

Figure 2. Configuration of microgrid (MG). (**a**) bidirectional energy flow between load controller and UG, (**b**) bidirectional energy flow between load controller and battery, (**c**) one directional energy flow from WG to load controller, (**d**) one directional energy flow from load controller to NS, (**e**) one directional energy flow from load controller to WH.

The microgrid described in this work can be connected to the energy hub. An energy hub is considered a unit where multiple energy carriers can be converted, conditioned, and stored [34]. It is to be noted that this microgrid output is very stochastic and has the best cooperation with the UG. Through the UG there is also a possible connection with other energy carriers.

3. Input Data

As input data, a time series of production of the WG which is scaled to rated power P_{nom} = 5 kW, manufactured by TUGE Ltd (TUGE Ltd., Paldiski, Estonia) [35] was used. It is located in a coastal area with coordinates N 59.087694, E 23.591719. The dataset collected covers the period from 1 December 2015 to 30 November 2016, taking into consideration that December is the first winter month, and facilitating further seasonal analysis. The average WG power output is derived by dividing the electricity generated during the last sampling period by the length of the sample period Δt. In the actual research, Δt = 5 min, and a year is divided into intervals of 5 min in length. Raw consumption data is measured at a frequency of 4 times per second with the network analyzer—Chauvin Arnaux (Chauvin Arnaux Metrix, Paris, France). Unfortunately, raw data have not survived. After the measurement, raw data have been averaged to 10 s of time series data. In the next step, data were converted to 5 min averaged interval data. Production data was processed from the 5 min averaging interval. Generation and loads are sampled equally.

Figure 3 shows the data for one week in December 2015 in 5 min periods of energy units. Battery and WH capacities are considered usable net values. Adding supercapacitors as levelling elements to decrease excess power from wind generator to batteries in our case is not used because of the small probability to increase maximum power allowed to batteries. Energy-related parameters such as WH temperature are not appraised, and neither is the energy necessary to reach WH minimal temperature of 55 °C, to avoid proliferating Legionella bacteria [36]. Operating temperature is assumed to have been attained, and the state-of-charge charge/discharge dependency is neglected.

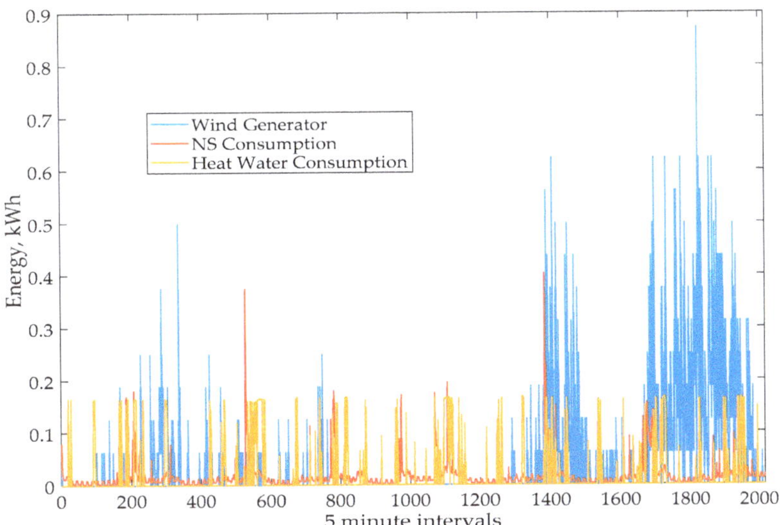

Figure 3. A week sample of initial data from December 2015.

The annual output of the WG is scaled to match the annual load as closely as possible. A typical private household's load pattern is considered with non-shiftable (NS) and hot water production (B) parts. The recorded annual electricity consumption was 3473 kWh, 47% (1632 kWh) being allocated to NS, and 53% (1841 kWh) to B, based on real measurement data [37]. We aimed to test the developed model and algorithm from our measured data. The Nordic climate conditions in Estonia are very changeable and even harsh given the very cold weather conditions in winter. Climate conditions are also accompanied by consumption data.

4. Simulation Model

In our simulation model, we first consider the control of energy flows in the power supply system of a private house that is sourced from a wind generator and that also has access to a utility grid. In Figure 2 we see the configuration of the microgrid (MG). The load controller in the center regulates sharing of energy produced by the generator.

This model has been created based on the principle that has been introduced in the sources [18,38]. This model belongs to multi-period multiple time scales over the year type [39]. In selecting wind turbine production data, we have based our analysis on one-year data with average wind conditions. Thus, the results of the simulation do not reflect the results for the different years to be taken into account when using outputs. This methodology does not command forecasting or economic issues [40]. These topics are planned in the following studies.

The sharing algorithm in every time interval is as follows [41], the variables denote (all in kWh):

X_1: is the output energy of wind generator in the current interval.
X_2: is the energy stored in a battery in the current interval.
Z_1: is the energy needed for NS consumption in the current interval.
Z_2: is the energy needed for hot water equipment in the current interval.

The simulation algorithm is delineated as follows:

(1) Next time interval is opened. Main action—the rest of the energies from previous intervals is transferred to the current one.
(2) WG output X_1 first satisfies NS loads Z_1.

(3) If $X_1 > Z_1$, the quantity $X_1–Z_1$ forwarded to WH. Steps 4 and 5 are skipped.
(4) If $X_1 < Z_1$, the quantity $Z_1–X_1$ is taken from the battery, if possible, i.e., if $Z_1–X_1 < X_2$. Otherwise, missing energy comes from UG.
(5) Energy Z_2 is taken from a battery or UG.
(6) Energy movements saved.

If the process does not require all the energy produced in the current time interval, the excess will move to the battery, and if there is still more energy left after that, it will go to the utility grid.

For evaluation of effectiveness and distinguishing these configurations and energy flows, this paper introduced and used a new cover factor. In literature, one can find several forms of cover factors. In our numerical experiments, expressed in detail below, the following formula was applied:

$$Y_D = (W_1 + W_2 + W_3 + W_4)/W_{total} \tag{1}$$

where:

1. Y_D is the demand cover factor.
2. W_1 is the total annual amount of energy produced by WG, which is directly consumed by NS devices needs.
3. W_2 is the total annual amount of energy produced by WG, which is used with hot water consumption from WH.
4. W_3 is the total annual amount of energy produced by WG, which is used for NS consumption from the battery during the year. If the WG is unable to supply directly to NS load, the missing energy will be taken from the battery.
5. W_4 is the total annual amount of energy flowing back from BB to MG. This is if there is not enough power in the battery and WH, then the energy is taken from BB. This is new in this paper. The flow is depicted as Arrow D in Figure 1b and is decisive for finding the new demand cover factor value Y_D. If there is not enough energy in BB, it is taken directly from UG Arrow C on Figure 1b.
6. W_{total} is the total annual amount of energy consumed for water heating and non-shiftable load, in other words, the total energy consumption of the entire household.

The cover factor, shortly, is the ratio of energy produced by the wind generator, which is consumed in the household under consideration. It is easy to see that the following inequalities hold:

$$0 \le Y_D \le 1 \tag{2}$$

Indeed, $Y_D = 0$ only when the wind generator exceptionally does not produce anything during the year. Denote with W the total annual amount of energy produced by the wind generator. For batteries in microgrid and BB, one can only load energy from a wind generator. It means that energy flow through Arrow E on Figure 1b can be only from the wind generator. The same holds for Arrow D and the same fact is true for MG battery. Y_D has no unit.

We have, therefore, the following expression:

$$W_1 + W_2 + W_3 + W_4 \le W \tag{3}$$

Because the energy amounts W_1, W_2, W_3, and W_4 used for microgrid consumption loads are produced by the same WG during the year. Mention that the initial conditions for energy in storage devices are set to zero in simulations. As the consumption from WG cannot exceed the total consumption inside MG then:

$$W_1 + W_2 + W_3 + W_4 \le W_{total} \tag{4}$$

The last inequality holds because the total consumption load W_{total} may contain a part of energy acquired from the utility grid as we see from Steps 4 and 5 of the algorithm.

According to the definition based on Equation (1) of cover factor Y_D, calculation from Equation (4) concludes that $Y_D \leq 1$. Therefore, the inequalities in Equation (2) are proved.

The microgrid solution we offer, together with the BB between MG and UG, is subject to certain limitations. This solution is designed to meet the needs of a private house as it is detailed in the above described methodology. This system is built on the principle that the uncertain parameters of the WG output power are grounded in storage devices such as WH and batteries. The present approach adopted in this work is mainly focused on component- and system-level design approaches rather than taking system parameter uncertainty modelling as done by the source [42] or by using robust optimal energy management [40]. There is no way to include electric car chargers in this system, nor, for example, fast boilers. To do this, the microgrid must be built differently, given the larger instant consumer power.

5. Results and Discussions

To achieve the goals of this work, computer simulations were carried out. The whole energy system of the household Figures 1 and 2 is inserted into a simulation model with linear charge and discharge characteristics. The input variables for the model are the year-long time series of wind generator output; non-shiftable loads; hot water consumption. The length of all three time series is 105, 120 and this is also the number of time intervals we used.

To estimate the coincidence between generation and load, the cover factor described by Equation (1) is applied. Cover factor Y_D characterizes the local generation/demand ratio. The simulation is based upon flow charts in Figure 1.

Based on Figure 4, the values of microgrid equipment parameters for numerical simulation experiments have been fixed. Because of practical considerations and expert assessment, the capacitance values of both the microgrid battery and WH are chosen to be 6 kWh. In future work, we intend to find a numerical method for the quantitative evaluation of these values and estimation of cover factor Y_D increments.

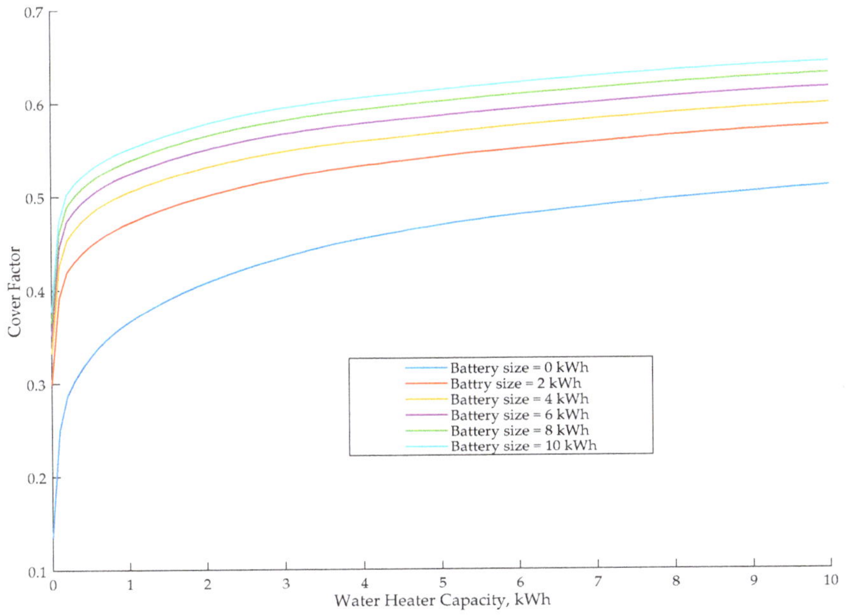

Figure 4. Cover factor Y_D dependency on battery and water heater (WH) capacity.

In the next stage of the simulation experiment, the amplification coefficient R_S was introduced to find the dependence in wind generator production of energy flow from microgrid through BB to utility grid Arrow F in Figure 1b and vice versa Arrow A in Figure 1a. The amplification coefficient (R_S) is the ratio of the energy produced per year by the wind generator to the energy consumed in the microgrid. With different values of coefficient R_S, the output of the wind generator is multiplied. Figure 4 depicts energy flows, where R_S varies between 0.8 and 3. At the crossing point or collocation point of two lines, R_S = 1, (it means the original production time series), the absorbed and injected energy become equal. Further increase of R_S results in a linearly growing part of generated electricity fed into the utility grid. The energy, absorbed from the UG, is characterized by a slightly falling line.

In equilibrium Figure 5, the exported and imported energy equal both 1401 kWh, with Y_D being = 0.597. Even a minor wind generator over-leverage results in significantly more electricity being injected into the utility grid.

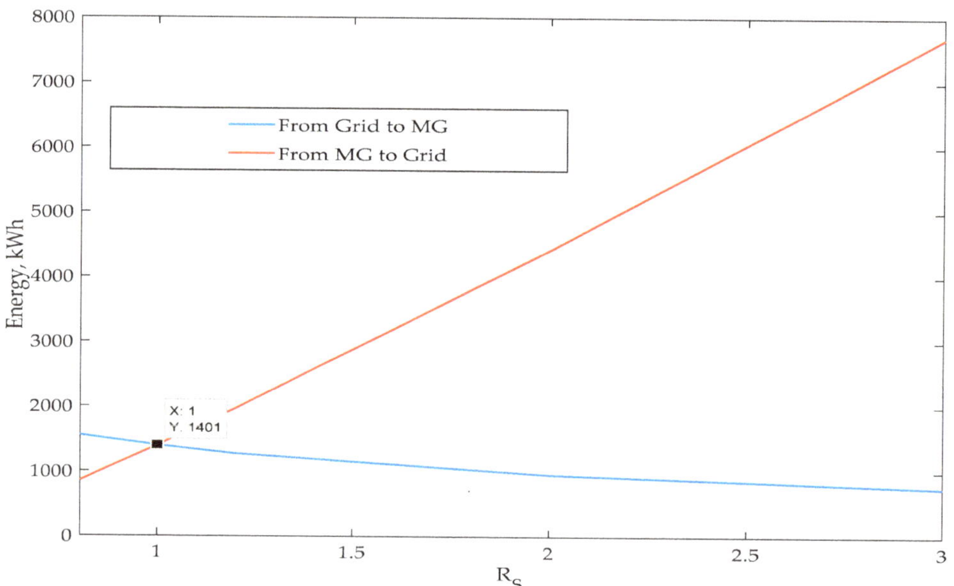

Figure 5. Dependence of energy flows between the utility grid (UG) and microgrid (MG) without buffer battery (BB) for different R_S values.

The next experiment examines the dependency of energy flows from the buffer battery size. Figure 6 shows the direct energy flow from grid to MG Figure 1b, Arrow C. Figure 6 depicts the electricity flow back from BB to MG Figure 1b Arrow D. Considering BB size 10 kWh in Figure 5, the grid-to-microgrid energy is equal to 1180 kWh, while BB-to-microgrid energy becomes equal to 221 kWh Figure 6. These two numbers sum up as 1401 kWh, which is valid for the "bufferless" case Figure 5. The same approach can be applied to other R_S and BB values.

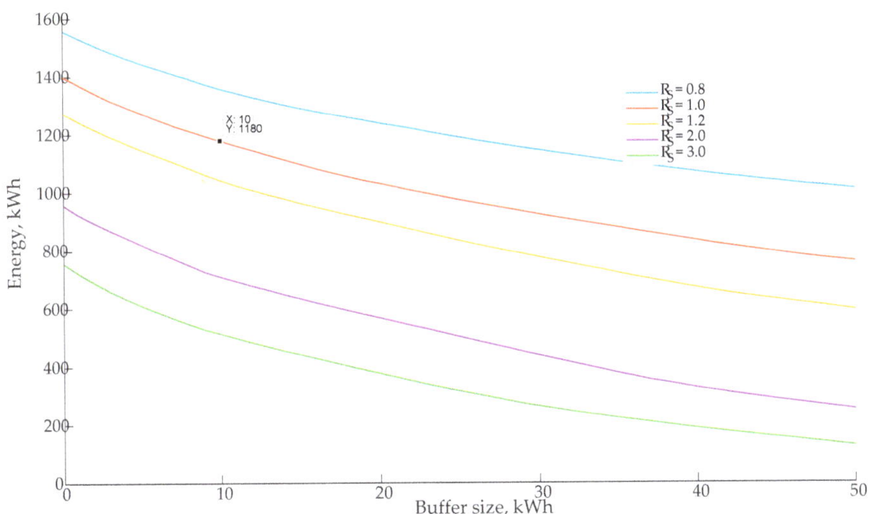

Figure 6. Energy from the utility grid (UG) to microgrid—MG.

Figure 6 shows direct energy flow from UG to MG Figure 1b, (C). Figure 7 depicts the electricity flowing back from BB to MG Figure 1b (D). Considering BB size 10 kWh Figure 5 the grid-to-microgrid energy is equal to 1180 kWh, while BB-to-microgrid energy becomes equal to 221 kWh Figure 5. These two numbers sum up as 1.401 kWh, which is valid for bufferless operation Figure 4. The same approach applies to other R_S and BB values as well. In Figure 7, energy amounts are in the y-axis on the Formula 2 fraction line the fourth member W_4.

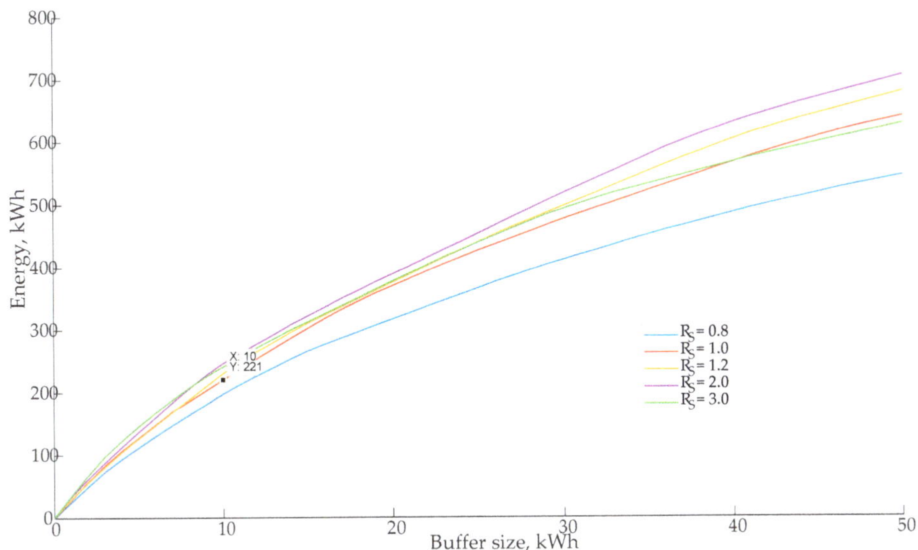

Figure 7. Energy from buffer battery (BB) to microgrid (MG).

Figure 8 depicts the energy flow from BB to UG. If $R_S = 1$ when moving to UG energy amount is 1164 kWh, which is nearly equal to MG entering energy flow, that is 1180 kWh, as in Figure 5. It was found that this difference is not decisive. The small difference is caused by $R_S = 1$ overproduction of 100 kWh from consumption, which is due to scaling conditions.

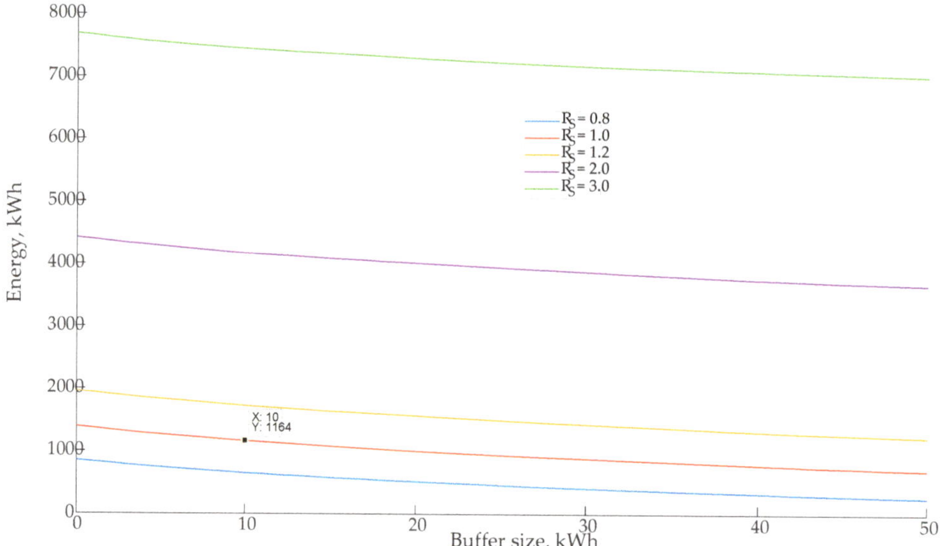

Figure 8. Energy from buffer battery (BB) to the grid.

Increasing R_S moves all residual energy to UG, as seen in Figure 8. In Figure 9, it can be seen dependence from BB size to cover factor by different R_S. The reasonable capacity of BB is 10 kWh. If BB = 0 when by $R_S = 1$ is $Y_D = 0.597$ and by BB = 10 kWh $Y_D = 0.66$. If we have $R_S = 2$, when $Y_D = 0.796$. As seen in Figure 9, the cover factor increase occurs when R_S is much bigger than the BB size, but BB size multiplies R_S influence.

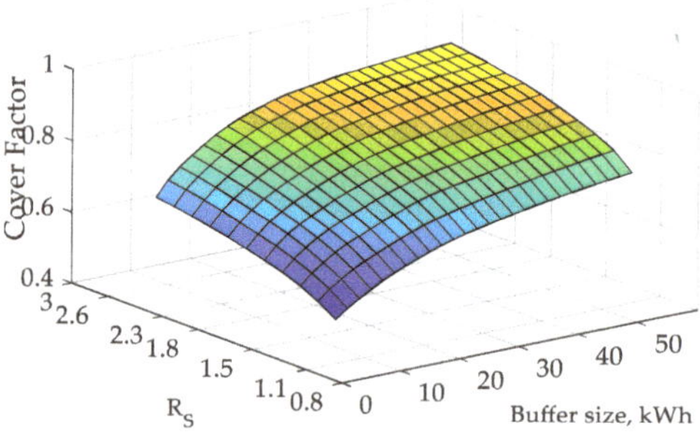

Figure 9. Influence of BB size and R_S on cover factor.

Based on Figure 9, results reveal that both, the BB and RS have an obvious influence on the demand cover factor. This suggests that the greater capacity factor of WG causes an additional increment in Y_D.

6. Conclusions

This paper developed and presented a simulation model of energy flows to investigate the self-consumption of a household with a wind generator as its independent source of electricity. The main assessment parameter is the demand cover factor (1). As input data, time series of typical private household load patterns with non-shiftable (NS) and hot water consumption (B) parts Figure 3, as well as the time series of production of the wind generator (WG) with rated power P_{nom} = 5 kW, were used. This simulation approach is easily generalized/replicated in more complex configurations of households, and also for different individual places such as warehouses, industrial buildings, etc. By increasing local consumption from renewable energy sources, energy losses in the utility grid and greenhouse gas emissions are reduced.

The study demonstrated that a buffer battery introduced between the utility grid and household inner microgrid has a notably positive influence on the demand cover factor. As a result, it is possible to reduce the amount of energy purchased from the utility grid, as seen in Figures 6 and 7. Figure 6 shows that if BB = 10 kWh, R_S = 1, is purchased from UG to MG 1180 kWh. If R_S = 2 and BB= 10 kWh, we only need to buy 701 kWh of electricity from UG. Wind turbine oversizing is reasonable based on these calculations, as wind production varies greatly over the years. Figure 7 shows the amounts showing how much less electricity we buy from UG compared to BB = 0.

It was proved that the production of the wind turbine is crucial for the value of the demand cover factor. To demonstrate this, the amplification coefficient R_S was introduced to find the dependency of wind generator production of energy flow from microgrid to the utility grid and vice versa. Figure 9 shows that compared to the baseline situation as R_S = 1 and BB = 0 Y_D = 0.597 then in a situation where R_S = 2 and BB = 10 kWh are Y_D = 0.796. This is a significant increase.

Moreover, the amplification coefficient R_S affects the cover factor only when applied to wind generator production. Numerical experiments showed that when applied to consumption data, the change is marginal.

Future research may focus on the WG production forecast and economic factors and parameters, which should be included in real simulation models.

Author Contributions: A.A. supervised the work and prepared the original research; W.Y. and M.L. provided the support in formulating the simulation problem; R.I. and T.K. draft preparation and evaluation of manuscript; P.M. conceptualization and MATLAB simulation. All authors have read and agreed to the published version of the manuscript.

Funding: This research was funded by the Estonian Centre of Excellence in Zero Energy and Resource Efficient Smart Buildings supported this research and Districts, ZEBE, grant TK146 funded by the European Regional Development Fund.

Acknowledgments: The authors would like to thank Tuge Energia Ltd., for making the data of the 10 kW wind generators available, Alo Allik for consumption data for this study and the Estonian Centre of Excellence in Zero Energy and Resource Efficient Smart Buildings and Districts, ZEBE, grant TK146 funded by the European Regional Development Fund supported this research.

Conflicts of Interest: The authors declare no conflict of interest.

References

1. Energy Efficiency Directive. Available online: https://ec.europa.eu/energy/topics/energy-efficiency/targets-directive-and-rules/energy-efficiency-directive_enstrategy (accessed on 10 January 2021).
2. Naval, N.; Sánchez, R.; Yusta, J.M. A virtual power plant optimal dispatch model with large and small-scale distributed renewable generation. *Renew. Energy* **2020**, *151*, 57–69. [CrossRef]
3. Annuk, A.; Jõgi, E.; Hovi, M.; Märss, M.; Uiga, J.; Hõimoja, H.; Peets, T.; Kalder, J.; Algirdas, J.; Allik, A. Increasing self electricity consumption by using double water heating tanks for residential net zero energy buildings. In Proceedings of the 6th International Conference on Renewable Energy Research and Application (ICRERA), San Diego, CA, USA, 5–8 December 2017; Volume 6, pp. 106–110. [CrossRef]
4. Chub, A.; Zdanowski, M.; Blinov, A.; Rabkowski, J. Evaluation of GaN HEMTs for high-voltage stage of isolated DC-DC converters. In Proceedings of the 10th International Conference on Compatibility, Power Electronics and Power Engineering (CPE-POWERENG), Bydgoszcz, Poland, 29 June–1 July 2016; pp. 375–379. [CrossRef]

5. Blinov, A.; Chub, A.; Vinnikov, D.; Rang, T. Feasibility study of Si and SiC MOSFETs in high-gain DC/DC converter for renewable energy applications. In Proceedings of the IECON 2013—39th Annual Conference of the IEEE Industrial Electronics Society, Vienna, Austria, 10–14 November 2013; pp. 5975–5978. [CrossRef]
6. Lepa, J.; Annuk, A.; Kokin, E.; Põder, V.; Jürjenson, K. Energy production and consumption charts in energy system. *Oil Shale* **2009**, *26*, 309–318. [CrossRef]
7. Zemug, L.; Su, S.; Yunning, Z.; Xiaolomg, J.; Houhe, C.; Yujiang, R. Energy management strategy of active distribution network with integrated distributed wind power and smart buildings. *IET Renew. Power Gener.* **2020**, *14*, 2255–2267. [CrossRef]
8. DeCicco, M.; Liu, D.Y.; Heo, J.; Krishnan, R.; Kurthen, A.; Wang, L. Carbon balance effects of U.S. biofuel production and use. *Clim. Change* **2016**, *138*, 667–680. [CrossRef]
9. Okok, M.O.; Ruchathi, G.; Oromat, E. Expanding access to clean energy in developing countries: The role of off-grid mini hydro power projects in Kenya. *Int. J. Renew. Energy Res.* **2019**, *9*, 1571–1577.
10. Hadiyanto, D.; Purwanto, Y.A.; Noorachmat, B.P.; Sapei, A. An indicator and evaluation criteria for off-grid micro-hydro power sustainability assessment. *Int. J. Renew. Energy Res.* **2019**, *9*, 1–15.
11. Zoss, T.; Karklina, I.; Blumberga, D. Power to gas and pumped hydro storage potential in Latvia. *Energy Procedia* **2016**, *95*, 528–535. [CrossRef]
12. Kesgin, M.G.; Han, P.; Taran, N.; Ionel, D.M. Overview of flywheel systems for renewable energy storage with a design study for high-speed axial-flux permanent-magnet machines. In Proceedings of the 8th International Conference of Renewable Energy Research and Applications (ICRERA) 2019, Brasov, Romania, 3–6 November 2019; pp. 1026–1031. [CrossRef]
13. Rehman, Z.; Mahmood, A.; Razaq, S.; Wamiq, A.; Naeem, U.; Shehzad, K. Prosumer based energy management and sharing in smart grid. *Renew. Sustain. Energy Rev.* **2018**, *82*, 1675–1684. [CrossRef]
14. Jie, C.; Bu, Z.; Wang, Y.; Yang, H.; Jiang, J.; Li, H.-J. Detecting prosumer-community groups in smart grids from the multiagent perspective. *IEEE Trans. Syst. Man Cybern. Syst. PP* **2019**, *99*, 1–13. [CrossRef]
15. Bautista-Villalon, M.; Gutirerrez-Villalobos, J.; Rivas-Araiza, E.A. IoT-based system to monitor and control household lighting and appliance power consumption and water demand. In Proceedings of the 7th International Conference of Renewable Energy Research and Applications (ICRERA) 2018, Paris, France, 14–17 October 2018; pp. 785–790.
16. Blinov, A.; Vinnikov, D.; Husev, O.; Chub, A. Experimental analysis of wide input voltage range qZS-derived push-pull DC/DC converter for PMSG-based wind turbines. In Proceedings of the PCIM Europe, Nuremberg, Germany, 14–16 May 2013; pp. 1435–1444.
17. Sambaiah, K.S. A review on optimal allocation and sizing techniques for DG in distribution systems. *Int. J. Renew. Energy Res.* **2018**, *8*, 1236–1256.
18. Vanhoudt, D.; Geysen, B.; Claessens, F.; Leemans, L.; Jespers, L.; Bael, J.B. An actively controlled residential heat pump: Potential on peak shaving and maximization of self-consumption of renewable energy. *Renew. Energy* **2014**, *63*, 531–543. [CrossRef]
19. Annuk, A.; Jõgi, E.; Lill, H.; Kalder, J.; Hovi, M.; Pihlap, H.; Härm, M.; Allik, A. Augmentation of self-consumption of electricity by using boilers and batteries for residential buildings. In Proceedings of the 7th International Conference of Renewable Energy Research and Applications (ICRERA), Paris, France, 14–17 October 2018; Volume 7, pp. 256–260. [CrossRef]
20. Budes, E.E.B.; Ochoa, G.V.; Obregon, L.G.; Arango-Manrique, A.A.; Alvarez, J.R.N. Energy, economic, and environmental evaluation of a proposed solar-wind power on-grid system using HOMER Pro®: A case study in Colombia. *Energies* **2020**, *13*, 1662. [CrossRef]
21. Katsaprakakis, D.A.; Zidianakis, G.; Yiannakoudakis, Y.; Manioudakis, E.; Dakanali, I.; Kanouras, S. Working on buildings' energy performance upgrade in mediterranean climate. *Energies* **2020**, *13*, 2159. [CrossRef]
22. Hossain, E.; Zawad, M.; Rakibul, K.H.; Islam, K.H.R.; Akash, M.Q. Design a novel controller for stability analysis of microgrid by managing controllable load using load shaving and load shifting techniques; and optimizing cost analysis for energy storage system. *Int. J. Renew. Energy Res.* **2016**, *6*, 772–786.
23. Eltamaly, A.M.; Mohamed, M.A.; Alolah, A.I. A novel smart grid theory for optimal sizing of hybrid renewable energy systems. *Sol. Energy* **2016**, *124*, 26–38. [CrossRef]
24. Hassanzadeh, M.N.; Fotuhi-Firuzabad, M.; Safdarian, A. Wind energy penetration with load shifting from the system well-being viewpoint. *Int. J. Renew. Energy Res.* **2017**, *7*, 977–987.
25. Duerr, S.; Ababei, C.; Ionel, D.M. Load balancing with energy storage systems based on co-simulation of multiple smart buildings and distribution networks. In Proceedings of the 8th International Conference on Renewable Energy Research and Applications (ICRERA) 2017, San Diego, CA, USA, 5–8 December 2017; Volume 5, pp. 175–180.
26. Jonaitis, A.; Miliute, R.; Deveikis, T. Dynamic model of wind power balancing in hybrid power system. *Turk. J. Electr. Eng. Comput. Sci.* **2017**, *25*, 222–234. [CrossRef]
27. Bashir, A.A.; Kasmaei, M.P.; Safdarian, A.; Lehtonen, M. Matching of local load with on-site PV production in a grid-connected residential building. *Energies* **2018**, *11*, 2409. [CrossRef]
28. Degefa, M.Z.; Lehtonen, M.; McCulloch, M.; Nixon, K. Real-time matching of local generation and demand: The use of high resolution load modeling. In Proceedings of the IEEE PES Innovative Smart Grid Technologies Conference Europe, Ljubljana, Slovenia, 9–12 October 2016.
29. Hauer, I.; Balischewski, S.; Ziegler, C. Design and operation strategy for multi-use application of battery energy storage in wind farms. *J. Energy Storage* **2020**, *32*, 101572. [CrossRef]

30. Li, X.; Lim, M.K.; Ni, D.; Zhong, B.; Xiao, Z.; Hao, H. Sustainability or continuous damage: A behaviour study of prosumers´ electricity cinsumption after installing household disreibuted energy resources. *J. Clean. Prod.* **2020**, *264*, 121471. [CrossRef]
31. Rogeau, A.; Barbier, T.; Girard, R.; Kong, N. Evoluotion of electrical distribution grid sizing considering self-consunption of local renewable producton. In Proceedings of the 24th International Confeence & Exhibition on Electricity Distribution (CIRED), Glasgow, UK, 12–15 June 2017; pp. 1–5.
32. Dancker, J.; Götze, J.; Schultz, E.; Könneke, N.; Beyrau, F.; Wolter, M. Optimal design and operation of a CHP based district heating system including a heat storage and electrode boiler to increase self-consumption. In Proceedings of the 2019 IEEE Inovative Smart Grid Technologies Conference—Latin America (ISGT Latin America), Gramado, Brazil, 15–18 September 2019. [CrossRef]
33. Kroposki, B.; Basso, T.; DeBlasio, R. Microgrid standards and technologies. In Proceedings of the IEEE Power and Energy Society 2008 General Meeting: Conversion and Delivery of Electrical Energy in the 21st Century, PES, Pittsburg, PA, USA, 20–24 July 2008. [CrossRef]
34. Geidl, M.; Koeppel, G.; Favre-Perrod, P.; Klöckl, B.; Andresson, G.; Fröhlich, K. Energy hubs for the future. *IEEE Power Energy Mag.* **2007**, *5*, 24–30. [CrossRef]
35. TUGE 10 Technical Data, 10 kW. Tuge Energia Ltd., Paldiski, Estonia, 2020; pp. 4–6. Available online: http://www.tuge.ee/products/tuge10 (accessed on 10 January 2021).
36. Volker, S.; Kistemann, T. Field testing hot water temperature reduction as an energy-saving measure—Does the Legionella presence change in a clinic's plumbing system? *Environ. Technol.* **2015**, *36*, 2138–2147. [CrossRef]
37. Annuk, A.; Hovi, M.; Kalder, J.; Märss, M. Consumption and Wind Production Time Series Data. Estonian University of Life Sciences: Tartu, Estonia, 2020. Available online: https://doi.org/10.15159/eds.dt.20.02 (accessed on 10 January 2021).
38. Baetens, R.; De Coninck, R.; Roy, J.V.; Verbuggen, B.; Driese, J.; Helsen, L.; Saelens, D. Assessing electrical bottlenecks at feeder level for residential net zero-energy buildings by integrated systemsimulation. *Appl. Energy* **2012**, *96*, 74–83. [CrossRef]
39. Sperstad, I.B.; Korpås, M. Energy storage scheduling in distribution systems considering wind and photovoltaic generation uncertainties. *Energies* **2019**, *12*, 1231. [CrossRef]
40. Hossein, S.M.; Carli, R. A residential demand-side management strategy under nonlinear pricing based on robust model predictive control. In Proceedings of the 2019 IEEE International Conference on Systems, Man and Cybernetics (SMC), Bari, Italy, 6–9 October 2019; pp. 3243–3248. [CrossRef]
41. Annuk, A.; Hovi, M.; Kalder, J.; Kabanen, T.; Ilves, R.; Märss, M.; Martinkauppi, B.; Miidla, P. Methods for increasing shares of self-consumption in small PV solar energy applications. In Proceedings of the 9th International Conference on Renewable Energy Research and Applications (ICRERA) 2020, Glasgow, UK, 26–29 September 2020; Volume 5, pp. 184–188. [CrossRef]
42. Shi, R.; Li, S.; Zhang, P.; Lee, K. Integration of renewable energy sources and electric vehicles in V2G network with adjustable robust optimization. *Renew. Energy* **2020**, *153*, 1067–1080. [CrossRef]

MDPI
St. Alban-Anlage 66
4052 Basel
Switzerland
Tel. +41 61 683 77 34
Fax +41 61 302 89 18
www.mdpi.com

Energies Editorial Office
E-mail: energies@mdpi.com
www.mdpi.com/journal/energies

www.ingramcontent.com/pod-product-compliance
Lightning Source LLC
LaVergne TN
LVHW070902170726
843515LV00005B/692